HALIDES

CHEMISTRY, PHYSICAL PROPERTIES AND STRUCTURAL EFFECTS

Chemistry Research and Applications

Additional books in this series can be found on Nova's website
under the Series tab.

Additional e-books in this series can be found on Nova's website
under the e-book tab.

CHEMISTRY RESEARCH AND APPLICATIONS

HALIDES

CHEMISTRY, PHYSICAL PROPERTIES AND STRUCTURAL EFFECTS

JEAN LEFEBURE
EDITOR

New York

For permission to use material from this book please contact us:
Telephone 631-231-7269; Fax 631-231-8175
Web Site: http://www.novapublishers.com

NOTICE TO THE READER

LIBRARY OF CONGRESS CATALOGING-IN-PUBLICATION DATA

ISBN: 978-1-62417-947-1

Published by Nova Science Publishers, Inc. † New York

CONTENTS

PREFACE

This book discusses the chemistry, physical properties and structural effects of halides. Topics include the physical properties of alkali halide clusters and cluster-assembled materials; the synthesis of the nano-crystalline organic-inorganic composite nylon-6,6 nickel carbonate membrane; nucleophilic substitution and palladium-catalyzed coupling mechanisms of aromatic halides; dislocation pinned by a monovalent ion in various alkali halide crystals; and the nonlinear elastic properties and Gruneisen constants of alkali halide.

Chapter 1 – The study of alkali halide cluster and cluster-assembled materials has revealed that they present unique physical and chemical properties that are quite different from those of the corresponding bulk crystals. In particular, alkali (X) halide (Y) cluster ions are commonly observed in the forms $(XY)_n^{-1,0,+1}$, $(XY)_nX^{0,+1}$ and $(XY)_nY^{0,-1}$ and their properties vary as a function of their size, charge and nature of atomic constituents, i.e., their properties are intimately related to the X and Y constituent electronic densities. A detailed characterization of each member of a cluster ion series permits a better understanding of the cluster properties. Theoretical calculations can be very useful in predicting atomic and molecular properties or parameters which might be difficult to verify experimentally such as binding energies, ionization potentials, vibration frequencies, and fragmentation patterns of atomic and molecular clusters. They can also be helpful in the interpretation of experiments by providing details of the geometrical configurations and of the electronic density distribution. In the present chapter, the authors will mainly focus on experimental data of charged species emitted from alkali halide targets and theoretical predictions of their relative stability, fragmentation energies and charge distributions. The authors will cover small cluster systems (e.g., n = 1 - 5) to nanometric size clusters (e.g., n = 20 - 40).

Chapter 2 – In this study, the nano-crystalline organic-inorganic composite nylon-6,6 nickel carbonate membrane was synthesised,subsequent to which (a) the physico-chemical characteristics of the membrane was evaluatedby employing ATR-FTIR, SEM, EDX, TEM,TGA, XRD, and porosity measurements, and (b) membrane potential measurements were carried out using different concentrations of KCl, NaCl and LiCl 1:1 electrolyte solutions($2.5 \times 10^{-2} \leq c$ (M)$\leq 1 \times 10^{0}$).For the potential measurements, the high C_2 concentrations of the electrolyte were maintained on one side of the membrane and low C_1 concentration of the same electrolyte on the other side of the membrane, where the solution-concentration ratio $\left(C_2/C_1\right)$ was maintained at 10 and the membrane potentials found to be higher at low pH values. Also, the authors observed the positive membrane potentials for the composite membrane and is an indication of the membrane to be negatively charged and this potential found to be dependent of cation selectivity and is decreasing in the order LiCl>NaCl>KCl. At low electrolyte concentrations, the membrane potential found to be high and with increase of concentration, the membrane potential deceased. In addition to the potential measurements, the successful application of Teorell-Meyer and Sievers (TMS), Kobatake et al. and Nagasawa et al. theories were also employed to estimate the charge density of the membrane immersed in different electrolytes and it was found that the charge densities decreased in the order KCl>NaCl>LiCl for 1:1 electrolytes. The extended TMS theory was used to investigate the transference numbers, mobility ratio, distribution coefficients, charge effectiveness, perm selectivity and concentration of fixed charge for 1:1 electrolyte solutions.

Chapter 3 – Unactivated aromatic halides typically do not react with nucleophiles unless submitted to extremely drastic reaction conditions. The extremely low reactivities of aromatic halides can be attributed to their sTable aromatic π electron structure. But the unactivated aromatic halides can readily undergo radical nucleophilic substitution ($S_{RN}1$) and palladium-catalyzed reactions under relatively mild conditions. The remarkable enhancement of the aromatic halide reactivities has been attributed to their activation by electron transfer and/or palladium oxidative insertion. The palladium-catalyzed coupling of aromatic halides has been become as one of the most important chemical transformations for the construction of carbon-carbon chemical bonds in synthetic organic chemistry. The related reaction mechanisms and their synthetic applications will be discussed with an emphasis on the author's research interests.

Chapter 4 – The force-distance ($F(x)$) relation, which represents the model overcoming a weak obstacle by a dislocation with the help of thermal fluctuations, is described here for short-range obstacle less than about 10 atomic diameters. Furthermore, the three models (the square, the parabolic, and the triangular $F(x)$) taking account of the Friedel relation are investigated with respect to the relation between effective stress and temperature for KCl:Li$^+$ and KCl:Na$^+$ single crystals. On the three force-distance relations, the value of T_c is almost the same irrespective of the kind of force-distance relation for both the specimens, and the value of T_c (254$\sim$257 K) for KCl:Li$^+$ is obviously small in comparison with that (266$\sim$271 K) for KCl:Na$^+$. T_c is the critical temperature at which the effective stress becomes zero and a dislocation breaks away from the defects of cubic symmetry around Li$^+$ or Na$^+$ only with the help of thermal activation.

Chapter 5 – The second-order elastic constants (SOECs) and third-order elastic constants (TOECs) of the alkali halides (LiF, NaF, KF, LiCl, NaCl and KCl) with NaCl structure is presented using the density functional theory (DFT) and homogeneous deformation method. The k-points and the cutoff energy are given reliable verification while TOECs are concerned. From the nonlinear least-square fitting, the elastic constants are extracted from a polynomial fit to the calculated strain-energy data. Both the SOECs and TOECs are in agreement with the available experimental and previous results. The Lagrangian stress is changing linearly for small Lagrangian strain and it is found that the nonlinear effects play an important role while the finite strains are larger than approximately 0.025. The pressure derivatives of SOECs are calculated directly using the TOECs, which is used to discuss roughly the phase transition pressure of alkali halide. The Grüneisen constants of long-wavelength acoustic modes char- acterizing the anharmonic properties of alkali halide are also presented.

In: Halides
Editor: Jean Lefebure

Chapter 1

ALKALI HALIDE CLUSTERS: EXPERIMENT AND THEORY

Francisco Fernandez-Lima,[1] Enio F. da Silveira[2]
and Marco Antonio Chaer Nascimento[3]*
[1]Department of Chemistry and Biochemistry,
Florida International University, Miami, Florida, US
[2]Department of Physics, Pontifícia Universidade Católica,
Rio de Janeiro, Brazil
[3]Chemistry Institute, Universidade Federal do Rio de Janeiro,
Rio de Janeiro, Brazil

ABSTRACT

The study of alkali halide cluster and cluster-assembled materials has revealed that they present unique physical and chemical properties that are quite different from those of the corresponding bulk crystals. In particular, alkali (X) halide (Y) cluster ions are commonly observed in the forms $(XY)_n^{-1,0,+1}$, $(XY)_n X^{0,+1}$ and $(XY)_n Y^{0,-1}$ and their properties vary as a function of their size, charge and nature of atomic constituents, i.e., their properties are intimately related to the X and Y constituent electronic densities. A detailed characterization of each member of a

---

* Corresponding author: Francisco Fernandez-Lima, Assist. Prof. Florida International University, Department of Chemistry and Biochemistry, CP-312 11200 SW 8th Str. Miami, Florida 33199; Ph: 305-348-2037; Fax: 305-348-3772; E-mail: fernandf@fiu.edu.

cluster ion series permits a better understanding of the cluster properties. Theoretical calculations can be very useful in predicting atomic and molecular properties or parameters which might be difficult to verify experimentally such as binding energies, ionization potentials, vibration frequencies, and fragmentation patterns of atomic and molecular clusters. They can also be helpful in the interpretation of experiments by providing details of the geometrical configurations and of the electronic density distribution. In the present chapter, we will mainly focus on experimental data of charged species emitted from alkali halide targets and theoretical predictions of their relative stability, fragmentation energies and charge distributions. We will cover small cluster systems (e.g., n = 1 - 5) to nanometric size clusters (e.g., n = 20 - 40).

1. INTRODUCTION

Atom clustering has largely attracted the interest of theoretical and experimental researchers because of the unique properties of these emsembles. Cluster systems can be produced from a large variety of materials and can vary in composition from single elements to molecular constituents. [1]

In general, clusters containing single elements have a strong tendency to maximize the number of bonds per atom, leading to a transition from planar (2D) to three-dimensional (3D) structures at small cluster sizes (n < 10 *e.g.,* gold [2-4], silver [5-7] and boron [8] clusters). For molecular clusters, as the number of molecular constituents increases, this 2D - 3D transition tends to occur for even smaller cluster sizes (n = 3 - 4), like those observed in hydrogen bonded clusters of polar molecules (*e.g.,* water [9-11] and ammonia [12, 13]). A second transition (macro3D) occurs in 3D space when certain geometries (meso-structures) stabilize the entire complex. For example, the geometry of carbon clusters can change from linear to rings and then to fullerenes as the cluster size increases. [14-17]

Over the last few years there has been increasing interest in studying alkali halide systems at the nanometer scale, mainly because of their special electrical, magnetic and optical properties. [18] Alkali halide clusters and cluster-assembled materials have shown unique physical and chemical properties, which are quite different from those of the corresponding bulk crystals.

In the present chapter, we will mainly focus on experimental data of charged species emitted from alkali halide targets and theoretical predictions of their relative stability, fragmentation energies and charge distributions. For some species, neutral clusters will also be considered. We will cover from small cluster systems (e.g., n = 1 - 5) to nanometric size clusters (e.g., n = 20 - 40).

2. EXPERIMENTAL ANALYSIS OF ALKALI HALIDE CLUSTERS USING MASS SPECTROMETRY

2.1. Cluster Ion Production: Flowing Gas Condensation, Atomic/Ionic/Cluster Bombardment, Laser Desorption and Electrospray Techniques

The heating of an XY alkali halide solid may cause the formation of X^+, Y^-, XY^0, $(XY)_2^0$ and X_nY^+ species, but rarely the desorption of larger clusters. [19, 20] The emission of large clusters (e.g., $(XY)_nY^+$ or $(XY)_nX^-$) requires a local high energy density deposition near the surface, usually obtained by energetic (keV / MeV/ GeV) particle impact (atom, molecule, cluster or ion) [21-23] or by intense laser irradiation (laser vaporization/ ionization / ablation) [24, 25]. Under different conditions, cluster ions can also be produced by inert-gas condensation techniques [26], electrospray ionization [27, 28] and field desorption [29].

A landmark in cluster beam generation occurred in 1973 when a flowing gas condensation source succeeded in producing alkali cluster beams. [30] Following this event, a number of studies on the production of relatively large alkali halide cluster ions by using keV Secondary Ion Mass Spectrometry (SIMS) were reported. [21, 31-37] In particular, mass/charge ratios up to 90 000 u/e were detected for $(CsI)_nCs^+$ cluster ions. [37] At the end of the 1980's, neutral heavy ion projectiles, typically keV Ar or Xe atoms, were used to produce cluster ions by impinging on an alkali halide salts (also known as the Fast Atom Bombardment technique). [38, 39] Alkali cluster ions have also been produced by chemical ionization (CI) with the assistance of a reagent gas (e.g., SF_6). [38]

Two other powerful techniques have been applied to the study of alkali halides since the early 1990's: Laser Induced Desorption (LID) and ElectroSpray Ionization (ESI). Cuboid base lattices were confirmed for NaCl, NaI, CsCl and CsI halides, from data obtained by laser vaporization of these salts. [24, 40] In particular, the LID method can generate cluster ions with over hundred members. Laser ablation studies on the $(XY)_nX^+$ and $(XY)_nY^-$ series of several alkali halides have shown that their desorption yields decrease as exp $(-k\,n)$, where $k \approx 2$ for both series; this suggests that the neutral component $(XY)_n$ plays an important role in the emission process (see Figure 1). [25]

A later modification to the laser ablation studies, called delayed extraction, increased the mass resolution of the time-of-flight mass spectrometers [41], and at the same time, permitted the study of the initial stages of cluster ion formation (e.g., $(CsI)_nCs^+$ [42, 43]) by modifying the ion dynamics in the laser plume.

Other developments in the laser ablation technique have also allowed: i) the study of $(CsI)_3Cs^-$ isomerization with picosecond laser pulses [44], and ii) the measurement of the structural electric dipole of large alkali halide clusters by using the third and the fourth harmonics of two $Nd3^+$:YAG lasers (e.g., one to desorb and one to ionize the clusters) [45].

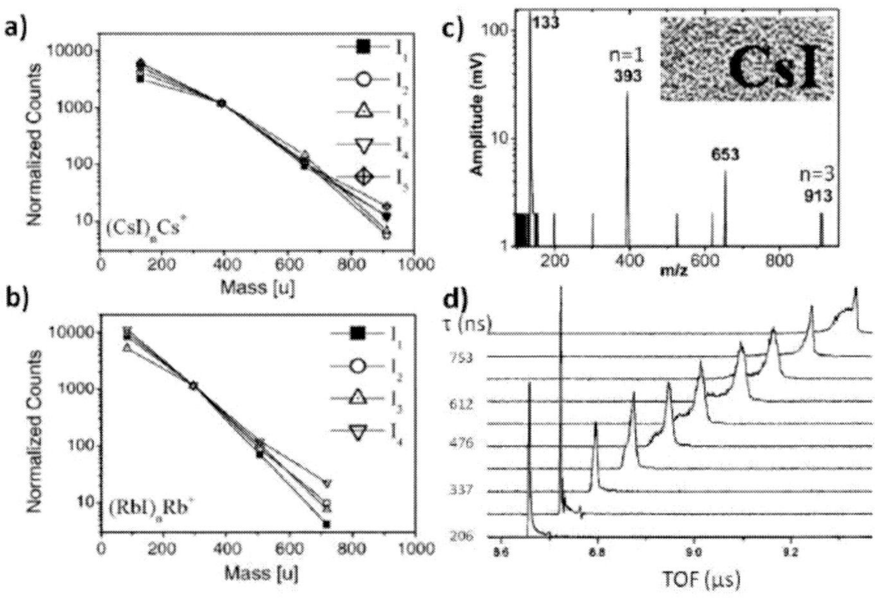

Figure 1. a) and b) Relative abundances (normalized to n = 1) of $(CsI)_nCs^+$ and $(RbI)_nRb^+$ positive ion clusters as a function of the laser intensity (I_{1-4} = 0.39-1.69 GW/cm^2); c) TOF spectrum of LID from a cesium iodide surface in positive ion mode (in the inset an optical image of the surface); and d) TOF distribution shape dependence on the delayed extraction (τ) parameter for the Cs^+ (m/z= 133) ion produced by LID.

Alternatively, ESI and MS/MS analysis have been employed to evaluate the influence of experimental parameters (e.g., solution concentration, pH, desolvation temperature, and solution flow rate) on the total positive cluster ion and negative cluster ion currents for several alkali halide salts. [46] ESI combined with ion trap mass spectrometry has also been used to study cluster structure and magic number effects for NaCl, NaF and other alkali halide cluster ions. [27, 47, 48]

2.2. Analysis of Cluster Ions Using Time-Of-Flight Mass Spectrometry: keV/ MeV Ion Impact and the ^{252}Cf Fission Fragment Source

The analysis of cluster ions using the time-of-flight (TOF) technique, a modality of mass spectrometry, can be performed in different operation modes: i) *event-by-event* [49], in contrast to the pulse mode, in which many ions with the same mass are emitted; ii) *coincidence counting* [50], in which coincidence among emitted ions with distinct masses is provided; iii) *delayed extraction* [51], in opposition to prompt or continuous extraction of the desorbed ions; iv) *ion-reflected mode* [52], where the flying ions are reflected by an electrostatic mirror, improving TOF resolution and/or metasTable ion decay analysis; v) *tandem* or *MS/MS configuration*, when the technique appears sequentially [53]. TOF applications involving alkali halide cluster ions can be found in $(LiF)_nLi^+$ cluster fragmentation [54], luminescence studies of self-trapped excitons in CsI [53, 55, 56], cluster-SIMS for analytical purposes and surface characterization [57, 58] and mainly in desorption mechanism studies [59-61].

When ion impact techniques have been used to produce cluster ions, the dependence of cluster emission yields on the local energy density has been mainly investigated in two conditions: i) high stopping power (high projectile charge states, projectile velocities close to the Bohr velocity, $V_B = 0.22$ cm/ns), and ii) non-linear stopping power (the desorption yield of a Z_n cluster projectile is higher than n times the desorption yield of the Z projectile) [62, 63]. From the former research line, the obvious conclusion is to perform bombardments with heavy ions by taking advantage of the highest charge state compatible with the desired beam flux. Easy to access projectiles for this purpose are fission fragments from a ^{252}Cf radioactive source (half-life of 2.6 years), emitted typically with the velocity of ~ 5 V_B (technique known as Plasma Desorption Mass Spectrometry, PDMS). The PDMS technique has been used to produce $(CsI)_nCs^+$ [59, 64-66], $(CsI)_nI^-$ [53], $(LiF)_nLi^+$ [62-64], $(LiF)_nF^-$ [62, 64] and other [22, 51, 53, 66] alkali halide cluster ion series. Although the secondary cluster ion emission of alkali halides can be easily obtained by this method, most of the studies have used LiF and CsI targets because they represent one of the lightest and heaviest species of the alkali halides, respectively. On the one hand, LiF cluster members have a relatively low number of electrons, a very important parameter for *ab initio* calculations; on the other hand, present a large cluster series, a convenient situation for studying the secondary ion mass distribution. Furthermore, since ^7LiF is commercially available and ^{19}F is the only fluorine sTable isotope, the LiF cluster series can be studied without isotope mixing. The CsI members are composed by high-Z elements transporting large number of electrons: their impact on solid surfaces transfers efficiently kinetic energy to targets and induces intense ion desorption, a desired quality for Secondary Ion Mass Spectrometry (SIMS). Both Cs and I have just one sTable isotope and, again, their cluster series are mono-isotopic and are commonly used for mass calibration over large mass ranges in mass spectrometers.

Reviews of early measurements of alkali halide electronic sputtering have been published in the period 1986 – 1994 [67-69]. The energies and desorption yields of secondary ions ejected from several alkali halides bombarded by a nitrogen beam in the 0.2-6 MeV energy range were studied in 1997. [61] Over the three following years, the ion induced desorption from LiF was intensely investigated by the TOF technique in a series of experiments focused on: i) the transition from the nuclear to the electronic sputtering regime [70], ii) the Li enrichment of LiF surface by ion bombardment [71, 72], iii) "axial" energy distribution of Li^+ and F^- desorbed ions [73], and iv) effects on the ion emission phenomena due to the projectile charge, clustering or velocity [74-77]. Complementary studies on LiF surfaces bombarded by heavy fast ions have been achieved by other techniques such as IR absorption, SEM and SAXS [78] and optical spectroscopy of color centers [79]. These studies have improved our knowledge on the formation of color centers, nuclear tracks, craters and hillocks/swelling by MeV projectile impacts on alkali halides, as well as our understanding on how clusters can be emitted from these materials. Experimental data for cluster ions, particularly for the first positive member of the series, are found in reference [75] for several alkali halide targets, and for LiF in refs. [70, 71, 74]. Findings on the desorption yield Y(n) dependence on the cluster size n for $(LiF)_nLi^+$ can be summarized as follows [23]: i) Y(n) increases with the projectile charge q as $Y(n) = Y_{00} + Y_0(n)\, q^5$, where Y_{00} is the desorption yield for a neutral beam; and ii) the function $Y_0(n)$ follows roughly $[S_e(E)]^{\alpha}$, where S_e is the stopping power and α a parameter of the order of 2-4.

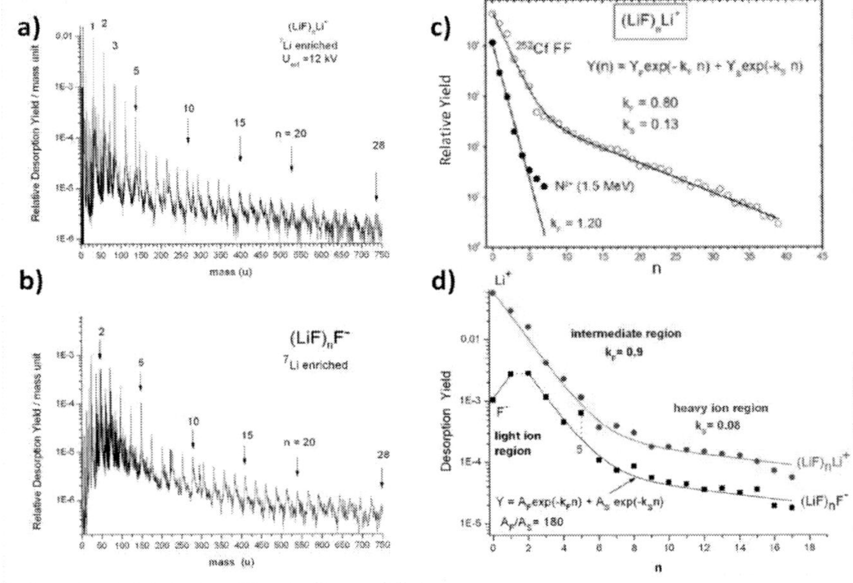

Figure 2. a) and b) PDMS positive and negative mode mass spectrum of $(LiF)_nLi^+$ and $(LiF)_nF^-$ cluster ions; c) Relative yield of $(LiF)_nLi^+$ ions emitted by the impact of ^{252}Cf fission fragments and of MeV nitrogen projectiles; (d) Desorption yields of $(LiF)_nLi^+$ and $(LiF)_nF^-$ ions emitted by the impact of ^{252}Cf fission fragments [23].

In addition, as shown in Figure 2, $Y_0(n)$ for positive and negative ions follows the empirical expression:

$$Y(n) = YF \exp(-kF\, n) + YS \exp(-kS\, n). \tag{1}$$

2.3. Angular Distribution and Initial Energy Analyses Using XY-TOF

Two-dimensional position-sensitive (*XY*) ion detectors can be combined with time-of-flight (TOF) instrumentation to set-up the XY-TOF technique for angular distribution and initial energy analyses of the desorbed ions. [80, 81] Briefly, the longitudinal (axial) component of the initial velocity is measured from the TOF value, while the XY part is responsible for the x and y transversal component measurements. Using these three components, the velocity modulus and the energy can be determined for each detected ion. The system also allows: i) improvement of the spectrometer mass resolution, ii) electric field mapping of ion flight region, and iii) measurements of the initial energy [82, 83].

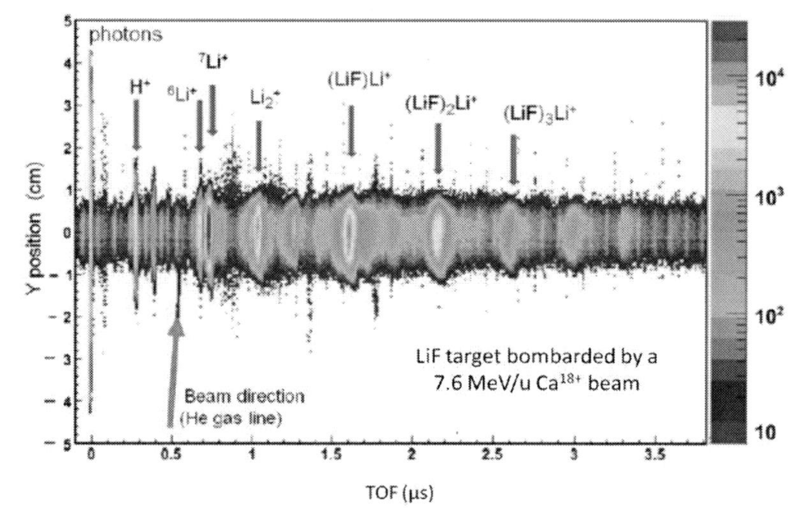

Figure 3. Three dimensional plot showing the desorbed positive ion events distributed over a *position* versus *time-of-flight* (Y-TOF) spectrum (more details in reference [84]).

The XY-TOF technique has been employed to study the cluster ion emission of LiF bombarded by neutral 1.0 MeV Ar [85], 7.6 MeV/u Ca^{18+} [84], 9.1 MeV/u Ca^{17+} [84], and 10.1 MeV/u (788 MeV) K^{33+} beams [86]. A Y(TOF) plot corresponding to a cleaved LiF mono-crystal target bombarded by 7.6 MeV/u Ca^{18+} at a 60° impact angle with respect to the surface normal is shown in Figure 3. The TOF values of the $(LiF)_nLi^+$ cluster ions are marked; note also the presence of the Li_2^+ and Li_3^+ metal cluster ions [84]. The results in reference [86] show that monomers and clusters have angular distributions symmetrical with respect to the surface normal to the target. The ejection of $(LiF)_nLi^+$ secondary ions is preferentially perpendicular to the LiF target and can be described by the $N(\theta) = A \cos^m(\theta)$ function, where $m \sim 4.2$ for Li^+ (n = 0) and decreases to m ~ 1 for n = 7 (the angular distribution becomes broader for heavier clusters). This behavior is quite distinct from the neutral particle emission, for which a jet-like structure has been reported. [86, 87] The analysis of the energy distribution of $(LiF)_nLi^+$ clusters has been treated in more detail in reference [86].

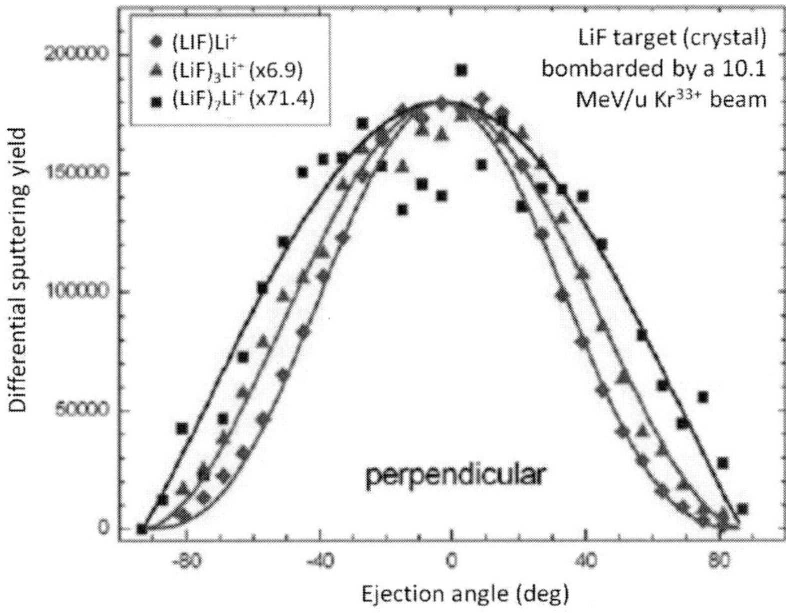

Figure 4. Angular distribution of $(LiF)_{n=1,3 \text{ and } 7}Li^+$ cluster ions from a LiF target bombarded by 101.1 MeV/u Kr^{33+} beam. Notice that the lighter the cluster the wider the angular distribution is in a plane perpendicular to the incidence plane [86].

2.4. Structural Analysis Using Ion Mobility Spectrometry – Mass Spectrometry

Despite the progress achieved in the characterization of alkali halide cluster ions using mass spectrometry, this experimental approach is limited in that it does not provide direct information on the cluster ion geometry or structure. To this end, the use of Ion Mobility Spectrometry (IMS) and more recently it's coupling to mass spectrometry have permitted the study of the gas-phase structure of alkali halide cluster ions. [88-90]

The principles for IMS separation have been described in detail elsewhere. [91-93] Briefly, ions enter a drift cell filled with an inert gas and migrate along the length of the drift cell under the influence of a weak electrostatic field. The ions experience collisions with the neutral drift gas establishing a constant drift velocity that is related to the electric field (E) through the following equation:

$$v_d = K \cdot E$$

(2)

where v_d is the drift velocity of the ion and K is a mobility constant that is specific to the ion. The mobility of an ion (K) is inversely related to its ion-neutral collision cross-section (the apparent surface area of the ion) and forms the basis for typical IMS separations; that is, a separation based on size, composition, structure, and charge state.

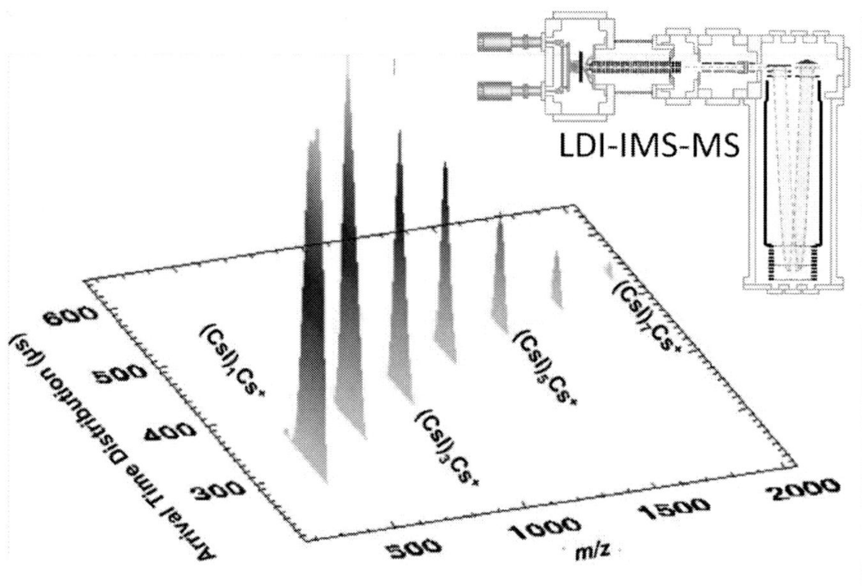

Figure 5. Three-dimensional IM-MS plot of $(CsI)_{n=1-7}Cs^+$ cluster ions produced by 355 nm laser desorption ionization. In the inset, schematic of a LDI-IMS-oTOF-MS instrument used in these studies.

Over the last decade, there has been an increasing use of Ion Mobility Spectrometry (IMS) for a wide variety of applications (e.g., see review [93]). This is partially caused by the developments of new types of IMS analyzers (e.g., periodic focusing DC ion guide, [94, 95] segmented quadrupole drift cell, [96] multistage IMS, [97] field asymmetric IMS [98], transient wave ion guide [99] and trap ion mobility spectrometers [100, 101]). A common pursuit has been to increase the mobility separation and ion transmission. Several groups have shown the advantage of coupling IMS to mass spectrometry (MS), thus achieving two dimensional separations based on the ion-neutral collision cross section (Ω/z) and the mass-to-charge, respectively. [102-105]

In particular, IMS-MS has the capability of separating ions of different classes along characteristic mobility trend lines (e.g., fullerenes, peptides, nucleotides, lipids, etc.); the IMS-MS separation in chemical classes is of great utility in the analysis of complex mixtures in the field of proteomics, [106, 107] glycomics, [108] metabolomics, [109] and petroleomics [110].

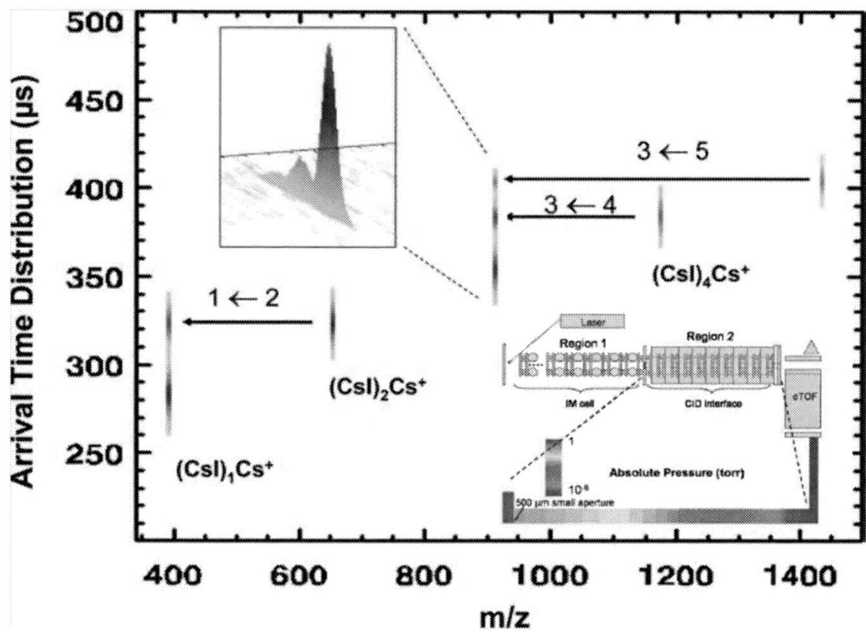

Figure 6. Two-dimensional IM-MS projection plot of the $(CsI)_nCs^+$ CID channels observed at a field strength of 40 V/cm in the CID interface (region 2). A clear separation of the parent (laser desorbed) and fragment (IM-CID-MS) ion signals is observed. In the inset; bottom) schematic of the CID interface electrode geometry and pressure gradient (region 2), and top) three-dimensional IM-MS plot of the ion signal observed at m/z = 913 corresponding to the n = 3 parent ion and the 4->3 and 5-> 3 dissociation channels.

Combined with theoretical calculations of the conformational space of molecular ions, IMS permits the determination of candidate structures that give the best description of a given molecular systems (e.g., electronic states, clusters, peptides, and protein complexes). [90, 111, 112] In the case of alkali halide cluster ions, IMS has been used to measure the mobility of different cluster ions (see example in Figure 5). More recently, cluster ion mobilities and collision cross sections have been reported for the case of cesium iodide cluster ions. [90]

Studies have also shown that IMS-MS coupled with Collision Induced Dissociation (IMS-CID-MS) permits the simultaneous study of the relative stability of specific ion packages; that is, it has the unique advantage that pre-selection of the parent ions is not necessary, since they are already separated in the IMS space. [113, 114] For example, IMS-CID-MS can be used to study the dissociation channels and relative stability of alkali halide cluster ions (see example in Figure 6). [90] In particular, IMS-CID-MS permits simultaneous detection and analysis of different cluster sizes and isomers in the same experiment, which provides a way of studying the relative stability of a given structure as a function of the internal energy (or effective temperature in the IMS experiment).

3. THEORETICAL METHODS FOR STRUCTURAL CHARACTERIZATION OF ALKALI HALIDE CLUSTERS

From a theoretical point of view, alkali halide clusters are good candidates to evaluate the accuracy of theoretical predictions due to their elementary electronic and ionic bonding nature. Theoretical calculations based on phenomenological pair potential models have been successful in describing the main characteristics of an alkali halide cluster and can be very useful for finding local minima on the potential-energy surface. However, with the advent of predictive *ab initio* methods, a large number of alkali halide structures have been proposed. In particular, *ab initio* calculations including electron correlation have provided an accurate description of the most sTable candidate structures observed in gas-phase cluster ion experiments.

3.1. Genetic Algorithms As a Search Engine for STable Configurations

The search for small cluster geometries strongly depends on the initial guess structure regardless of the theoretical method. Over the last years, the use of genetic algorithms (GA) has improved candidate structures generation routines. A GA is an optimization and search technique based on the principle of natural evolution. [115] In particular, a GA does not include the calculations of derivatives and it has been successfully used in many areas of science and technology. [116-121] In the case of alkali halide cluster ions, the search for the global minimum using a GA is performed in the 3-N configurational space with the energy value as a criterion of fitness. [122] The individuals of the initial population are randomly generated, where the minimum inter-atomic distance and the maximum cluster volume are defined by the user as a function of the cluster size. A global geometry optimization of the cluster structures is obtained using GEGA [121] and all individuals are optimized to the nearest stationary point and the atoms coordinates are updated. Different isomers generated by the GA are stored in a "pool" of candidate structures. In each GA loop, a set of new individuals is created to replace the weakest individuals in the population. The percentage of individuals replaced is called GAP. For a consequent reproduction, each individual in the current population is assigned a certain probability to be chosen as a genitor according to its fitness. There are two main genetic operators responsible to create the new individuals: mutation and crossover. Crossover is the process through which the genetic information from two parent chromosome is combined when generating new individuals (e.g., cut and splice operator [120] or modifications of it). For example, in the crossover operator in ref. [122], each atom of the two different parents selected has an associated number. First, an orientation axis is randomly chosen among x, y, and z. Next, the atoms on each cluster are sorted with respect to this axis, from higher to lower coordinate points. In this case, the clusters are not cut by a plane, as in the crossover proposed in reference [120] but a cut point is chosen randomly in the selected axis, such as in a binary representation genetic algorithm. [115] The first offspring is formed by the upper or lower part (randomly selected) of the first parent. The remaining atoms are positioned as they appear on the second parent. The second offspring is formed in the same way but by changing the parents. The proposed crossover operator always generates clusters with the correct number of elements. To avoid stagnation and to maintain population diversity, a mutation operator that perturbs some of the atoms within the cluster is generally used. The cluster can be mutated by replacing the atomic coordinates of a given number of atoms with random values in two different ways. In the first, the atomic coordinates of some atoms are changed inside a sphere with defined radius around the atom. In this case the operator shifts the atomic coordinates. The radius of the sphere is selected by the user since it differs for each system being optimized. In the second way, the coordinates are randomly changed and the whole space is used. The mutation operator can completely change small structures such as alkali halide cluster ions (n < 15); however, the global geometry of small clusters is not so difficult to be found (normally on the first generations), and the operator is used to generate different local minima and maintain diversity.

To start a GA algorithm a number of initial parameters must be defined: types and number of atoms, charge and multiplicity, calculation method and basis set, interatomic distances, maximum cluster volume, number of generations, size of population, and crossover and mutation rates. Most of these parameters depend on the size of the cluster and are defined in a way that guarantees convergence (*i.e.,* the global minimum is obtained within few generations). The remaining generations contribute to the search for other low energy structures (isomers).

3.2. *Ab Initio* and DFT Methods for Geometry Optimization and Quantum Dynamics

The structures obtained from the GA algorithm must be refined using *ab intio* and/or Density Functional Theory (DFT) methods with appropriate basis sets. For the neutral and charged $(XY)_n$, $(XY)_nX$ $(XY)_nY$ clusters, geometry optimizations have been performed at the B3LYP/6-311+G(3df) and B3LYP/LACV3P*+ levels of calculation. [122-126] However, it is well-known that DFT calculations on charged species may be in error due to the fact that for most of the present available functionals the exchange energy does not exactly cancel the Coulombic self-interaction [127], but that this effect becomes less important as the number of atoms increases. To check for possible inconsistencies, the smaller structures (n= 2-10) have been also optimized at the MP2/LAVCV3P* level of calculation. For the neutral $(LiF)_{n=6,9}$ clusters, coupled-cluster calculations at the CCSD/SDD/ LAVCV3P* level have been also considered. From the results of the *ab intio* and the DFT calculations no substantial changes have been observed, either in the geometrical parameters or in the relative stability of the clusters. No symmetry restrictions have been imposed in the process of geometry optimization. A vibration frequency analysis has been performed for all the optimized structures at the level of calculation employed. For the reported structures, all frequencies have been observed to be real, indicating that the optimized structures correspond to the true minima in the respective potential energy hypersurfaces. The energy values of the optimized structures have been corrected for the zero-point energy (ZPE) to obtain the total energy E_T (E_T = SCF + ZPE). The nature of the charge distribution on the clusters has been investigated by computing the atomic charges using the CHelpG algorithm. [128] The DFT and MP2 calculations have been performed using the Jaguar 7.0 program [129] while for the CC the GAMESS/US-10 program has been used [130].

As will be discussed in the next section, neutral $(XY)_n$ clusters may exhibit quite sTable tube-like structures. In order to investigate their thermal stability, quantum dynamics Born-Oppenheimer calculations (NVT ensemble at T=300K) have been performed for the $(LiF)_{28}$ nanotube with octogonal cross-section, at the DFT/B3LYP-6-31G level of calculation. [131]

3.3. Alkali Halide Structures. *C(n,q,i)* Cluster Series

The following series of clusters have been considered: $(LiF)_n F^-$, $(LiF)_n Li^+$ and $(LiF)_n Li$ for n =1 - 9; $(LiF)_n$ for n = 1 - 36 and $(CsI)_n Cs^+$ for n = 1 - 8. The GA discussed in *Section 3.1* was applied to generate a pool of candidate structures for the smaller lithium-fluorine clusters (up to n = 5), while for the larger clusters the search was directed to specific series. For these clusters, as the number of isomers and variety of structures (families) increase with the size of the cluster, it is convenient to adopt a notation to label them according to their size and type of family. Thus, for the lithium-fluorine clusters each isomer structure is labeled as *C(n,q,i)*, where *n* is the cluster size (number of LiF units in the cluster), *q* is the cluster charge in atomic units (+1, 0 or -1) and *i* is the isomer index which characterizes the family type and/or the total energy: *i* = 1 for the linear structures, *i* = 2 for the planar rhombus family, *i* = 3 for regular polygons or cyclic planar shapes. Values of *i* ≥ 4 denote tri-dimensional (3D) structures: *i* = 4 for cyclic parallel polygons or pyramidal shapes, *i* = 5 for cubic or cube-like structures (tending to the fcc one) and *i* > 5 for others 3D shapes (chair, spherical-like, caravel, etc.).

Figure 7a) shows some *C(n,0,i)* structures obtained for the $(LiF)_n$ clusters classified according to the *C(n,q,i)* notation previously described. As the cluster size increases, the structures in the series change from planar to cube-like, while linear structures are observed for all cluster sizes. Figure 7b) shows the structures obtained for the $(LiF)_n Li^0$ (n = 1 − 9), identified by the same *C(n,0,i)* notation. Two linear $C_{Li}(1,0,1)$ structures are predicted at the B3LYP/LACV3P* level: Li-F-Li and Li-Li-F.

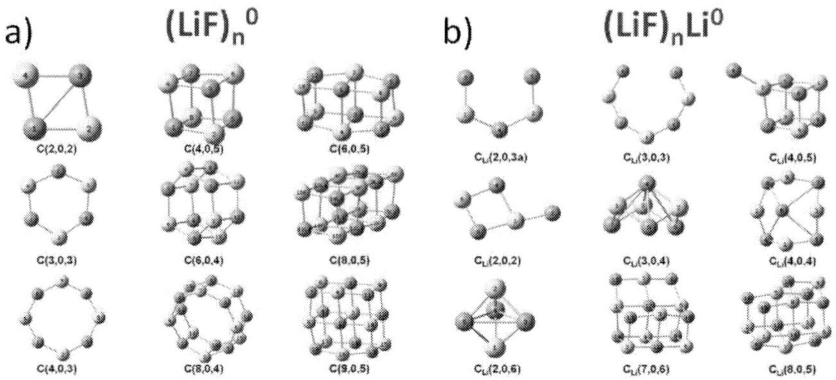

Figure 7 a) and b). Some predicted structures for $C(n,0,i)$ $(LiF)_n^0$ and $C_{Li}(n,0,i)$ $(LiF)_nLi^0$, respectively.

However, the former exhibit one imaginary frequency at the MP2/ LACV3P* level of calculation. The Li-Li-F structure is unique in the sense that no similar species were found for n>1. The planar families with $i = 2, 3$ and 4 have a common first member (a bent Li-F-Li) but the next members differ in the way the clusters grow. For n ≥ 4, cubic structures are also formed.

The shapes of the most representative $C_{Li}(n,1,i)$ $(LiF)_nLi^+$ structures are displayed in Figure 8a). For n=1, only the linear Li-F-Li$^+$ structure has been found to be sTable. The 2D configurations grow from a kite-like structure towards more complex kites/rhombic –like structures, or from a star-like structure. The 3D configurations are basically cubic-like, pyramidal, caravel-like and spherical (polyhedric-like). Figure 8b) shows some of the DFT structures found for the $(LiF)_nF^-$ series and classified according to the $C_F(n, -1, i)$ notation. Analogously to the positive $(LiF)_nLi^+$ series, only the linear F-Li-F$^-$ structure has been found to be sTable for the $(LiF)_nF^-$ series. The two-dimensional structures grow from either a kite-like structure toward more complex kite/rhombic -like structures or from a star-like structure. The three-dimensional configurations are basically cube-like, pyramidal and polyhedral-like. As the cluster size increases, various isomers have been also observed. However, not all the expected structures are observed, a result that reflects the fact that the $(LiF)_nF^-$ cluster ions are not only stabilized by short-range interactions, but by the average interaction of all the cluster counterparts. These findings are similar to those reported for the $(LiF)_nLi^+$, $(LiF)_nLi^0$ and $(LiF)_n^0$ series and may be a consequence of the ionic bonding nature of the LiF clusters.

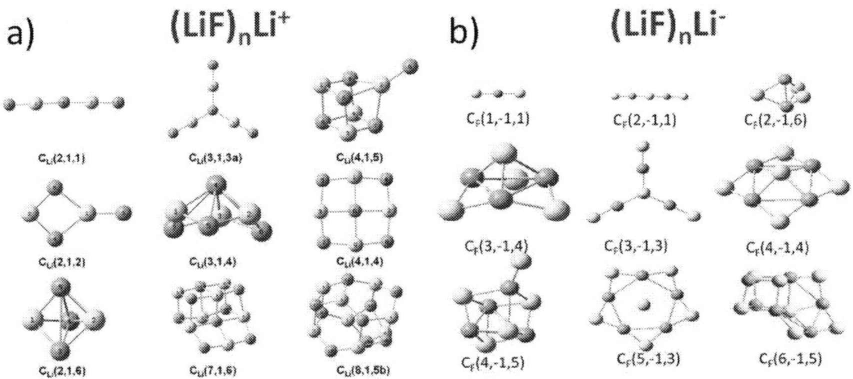

Figure 8 a) and b). Some predicted structures for $C_{Li}(n,1,i)$ (LiF)$_n$Li$^+$ and $C_F(n,-1,i)$ (LiF)$_n$F$^-$, respectively.

In the case of the (LiF)$_n^-$ series, only the linear isomers were considered, and following the same notation, this series can be denoted as $C(n,-1,1)$.

The fact that some of the (LiF)$_n$ tube-like structures show equal or greater stability than the isomeric cubic ones prompt us to further examine the members of these series. [131]

Figure 9 shows representative sTable members of the (LiF)$_n$ tube-like series with different cross-sections together with the equivalent ones for (NaCl)$_n$ and (KBr)$_n$ clusters.

Members of the LiF nanotube series exhibit tetragonal, hexagonal, octagonal, decagonal and dodecagonal cross-sections (Figure 9) and are classified according to the respective polygonal cross-sections (e.g., nanotube/tetragon).

The tetragon series would be, in principle, the one with the largest number of members, with a new member every two LiF units, the most sTable one corresponding to n = 8. The bond distance between the nearest Li and F atoms across the longitudinal section is $d_{Li-F} \sim 1.81$ Å, while the bond distances across the transversal central and terminal (or end-cup) sections are $d_{Li-F} \sim 1.96$ Å and $d_{Li-F} \sim 1.81$ Å, respectively. This trend of the transversal and terminal cross-sections is also observed for the other members of the series. The transversal cross section of the tetragon series presents an inner diameter of 0.241 nm (shortest diagonal) and lengths ranging from 0.183 to 1.279 nm.

The hexagon series starts at n = 6 and some of its members are the most sTable structures for a given cluster size, even when compared to the cubic ones, as will be discussed in the next section.

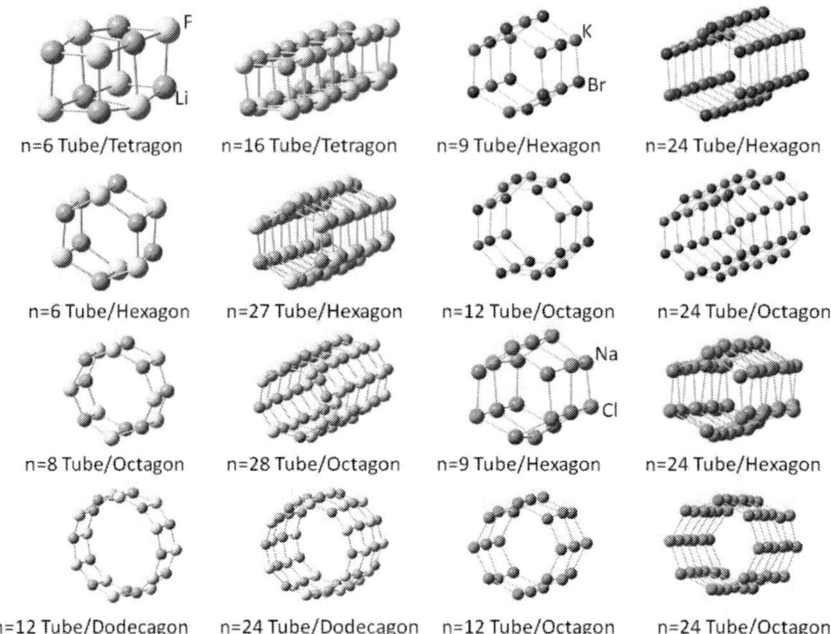

Figure 9. Representative $(LiF)_n$, $(KBr)_n$, and $(NaCl)_n$ neutral structures of the nanotube families.

For this series, the distance between the Li and F atoms across the longitudinal section is $d_{Li-F} \sim 1.86$ Å while across the transversal and terminal sections the distances are $d_{Li-F} \sim 1.96$ Å and $d_{Li-F} \sim 1.81$ Å, respectively. The transversal cross-sections present an inner diameter of 0.357 nm and lengths from 0.187 to 1.502 nm.

The octagon series starts at n = 8. The distances between the Li and F atoms across the longitudinal, the transversal and the terminal sections are $d_{Li-F} \sim 1.86$ Å, $d_{Li-F} \sim 1.93$ Å and $d_{Li-F} \sim 1.78$ Å, respectively. The transversal cross section of the octagon series presents an inner diameter of 0.472 nm and lengths from 0.188 to 1.138 nm.

The decagon and dodecagon follow the same pattern of all previously described nanotube series. The transversal cross section of the decagon and dodecagon tube series are 0.578 and 0.673 nm, respectively.

In particular, a similar pattern to that of the octagon and hexagon series was observed. For example, in the n = 24 dodecagon structure the Li and F atom distances across the longitudinal, transversal and terminal sections are $d_{Li-F} \sim 1.90$ Å, $d_{Li-F} \sim 1.90$ Å and $d_{Li-F} \sim 1.78$ Å, respectively.

These results also show that alkali halide nanotubes present much smaller diameters, ranging from 0.241nm to 0.67nm (LiF, tetragonal − KBr, hexagonal) when compared to the more traditional covalent nanotubes of C (1 - 2 nm), BN (1 - 3 nm) or GaN (30 - 200 nm). This is a consequence of the fact that the interactions are dominantly electrostatic and, therefore, practically no strain is involved in making ring-type structures differently from the covalent ones. Similar nanotube structures were obtained for NaCl and KBr halides at the DFT/B3LYP/LACV3P** level. [131] Representative structures of the hexagon and octagon series are shown in Figure 9. Comparisons between the different alkali halides show that the size of the structure scales with the atomic number from smaller (LiF) to medium (NaCl) to larger (KBr) structures, as expected if the constituents' typical bond distances are taken as references. For example, the transversal cross section of the nanotube/hexagon NaCl and KBr series presents a diameter of 0.523 and 0.644 nm, and lengths from 0.271 to 1.912 and from 3.301 to 2.312 nm, respectively. Analogously, the octagon NaCl and KBr series present a diameter of 0.654 and 0.847 nm, and lengths from 0.271 to 1.352 and from 3.301 to1.656 nm, respectively.

Figure 10 shows the sTable structures obtained for the $(CsI)_nCs^+$ clusters at the DFT/B3LYP/LACV3P** calculation level. [90] Structures labeled *I* are the most sTable ones among the isomers. Despite the extensive search for potential candidate structures, only three isomers for n = 2,5 and two isomers for n = 3,4,6,7 have been found to be real minima at the levels of calculation employed. Some of the structures previously proposed by Aguado et al. [132], based on an *ab-initio* perturbed ion model, present imaginary frequencies at both the B3LYP/ and MP2/LACV3P** levels of calculation. Moreover, several new structures were obtained in our search (*e.g.,* 3 II, 4 II, 6 II, 7 II and 8). For n > 3, some structures show similarities with the crystalline structure of the CsI bulk material, *e.g.,* structures 4 I, 4II, 5I, 5III, 6II and 7I still resemble a cubic structure.

3.4. Analysis of Alkali Halide Cluster Structures: Deviation Plot, Charge Distribution, Electron Affinity and Fragmentation Energy

The Deviation plot (D plot) methodology was initially proposed by the authors for a better and faster taxonomic description of the cluster isomers. [13, 16, 133, 134] The D plot analysis takes into account all the optimized structures from the "pool" of candidates regardless of the cluster size.

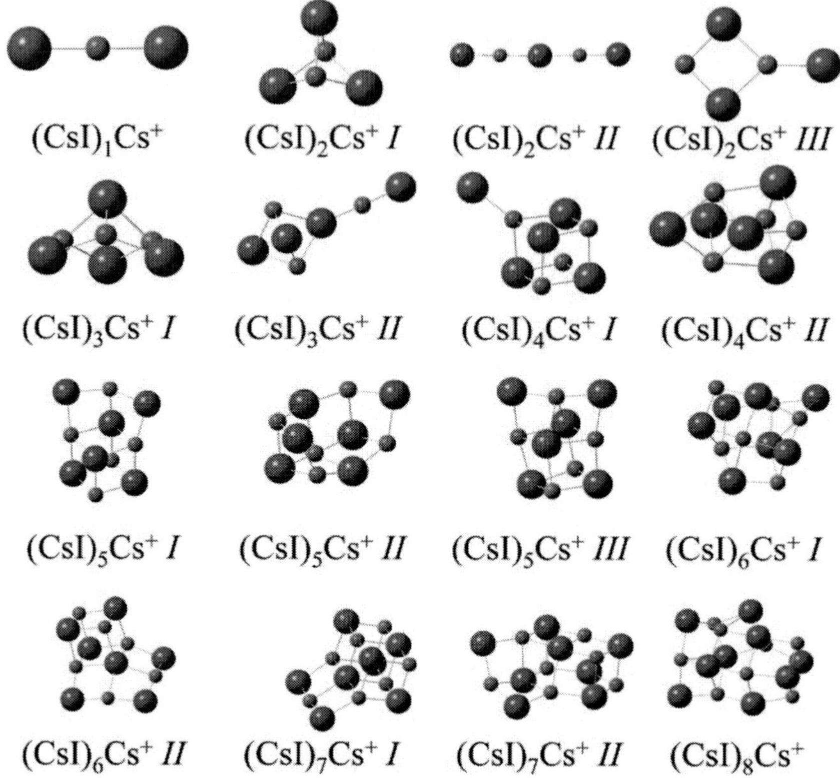

Figure 10. Optimized geometries of the $(CsI)_{n=1-8}Cs^+$ cluster ions at the DFT/B3LYP/LACV3P** calculation level.

The D plot is a practical method for displaying total energy differences of the distinct cluster configurations. It consists of trying to represent the total energy of a given cluster of size n and charge q as a function of the average energy of all of the n-isomers, $< E(n,q) >$, plus a certain energy deviation, D(n,q,i). Since, in general, the average energy $< E(n,q) >$ increases linearly with n, the total energy $E_T(n)$ can be expressed as:

$$E_T(n) = <E(n,q)> + D(n,q,i) = E_0 - (E_{XY})n + D(n,q,i) \quad (3),$$

where E_0 is the total energy of the X^+ or Y^- atomic ions and the slope coefficient E_{XY} is the average total energy of the neutral XY molecule. In practice, E_0 and E_{XY} are parameters determined from the linear fit over all predicted configurations. Since $D(n, q, i) << E_0 - E_{XY} \cdot n$, these two parameters are rather insensitive to the number of configurations employed, which is not the case for $D(n, q, i)$. Lower $D(n, q, i)$ values correspond to lower energy isomers and thus more sTable structures. The relevant features of the D-plot are that the most sTable isomers can be easily identified and the cluster families can be separated according to their trends in the D-plot.

An attempt to understand the relative stabilities of the clusters can be made based on their structures optimized as discussed in *Section* 3.2 and the respective atomic charges obtained from the ChelpG algorithm. [128] Inspection of the optimized structures shows a certain degree of symmetry, which could be related to the ionic bonding nature of the alkali halide clusters. The symmetry in the relative position of the atoms in the cluster is also observed in their atomic charge values.

In the case of 2D and 3D structures, the total cluster stability is mainly defined by multipole interactions. The effects of the symmetry and of the relative importance of monopole-dipole and dipole-dipole interactions for the cluster stability can be analyzed in the simpler scenario provided by the 1-D structures.

For the neutral $(LiF)_n$, since only one covalent, but highly polar bond, can be formed between the Li and F atoms, the LiF units of the cluster will be held together mainly by dipole-dipole type interactions which vary as $(1/r^3)$ where r is the distance between the dipoles.

For the linear structures, as the chain length increases the average distance between the LiF units increases while their dipoles do not change appreciably. Therefore, one should expect a decrease in the stability of the clusters with increasing the chain length as shown in the D-plot (Figure 11).

For the polygonal structures the dipole moment of the LiF units are identical to that of the single unit and the units are closer to each other than in the respective linear isomers.

Thus, one should expect the polygonal planar structures to be more sTable than the corresponding linear ones as observed in Figure 11.

On the other hand, as the number of units of the polygonal planar clusters increases the distance between the LiF units also increase and the stability decreases as observed in Figure 11. The same sort of reasoning can be used to analyze the relative stability of the 3D cyclic and cubic structures.

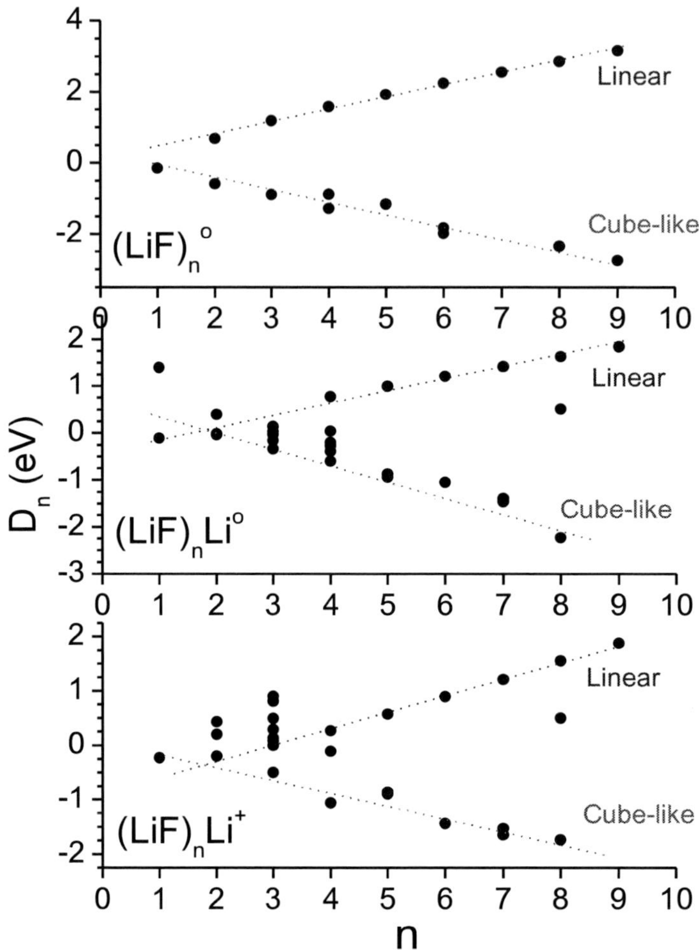

Figure 11. D plots for the $(LiF)_n^0$, $(LiF)_n Li^0$, and $(LiF)_n Li^+$ cluster series.

Quite interestingly, as the number of rings of the 3D cyclic structures increases, the number of first F neighbor atoms to a Li atom (or vice-versa) and the average distance among them may become similar to these properties in the traditional rocky salt cubic structure. Thus, for n>6, the 3D cyclic structures may become even more sTable than the cubic ones (Figure 11)

For the neutral $(LiF)_n F$ linear clusters the atomic charge distribution is symmetric relative to the central F (n odd) or Li (n even) atom and decreases, in absolute value, from the central atom to both ends of the chain. Thus, the clusters can be viewed as being formed from LiF units symmetrically displaced relative to the central Li atom (n odd) or to a central (Li-F-Li) unit for n even. The atomic charges in the LiF units are practically the same for all clusters and depend only on the position of the unit relative to the center of the chain. For this type of cluster, besides the dipole-dipole type interactions, for n even the cluster can be also stabilized by dipole-monopole interactions with the central charged Li atom, which vary as (q/r^2), where q is the charge of the ion. Due to the symmetric atomic charge distribution, for n odd, the dipole moment of the central (Li-F-Li) unit vanishes but not its charge. Thus, the interaction of the LiF units with the central (Li-F-Li) could be approximately described as a dipole-monopole interaction with the net charge of the central unit on the F atom. However, due to the extra dipole-monopole stabilization, the decrease in stability with increasing the chain length is not as pronounced as for $(LiF)_nLi$ cluster when compared to the $(LiF)_n$ ones (Figure 11). The cyclic planar structures for the $(LiF)_nLi$ clusters are not sTable because in order to form this type of structure one has to bring two positively charged Li atoms close together. For the same reason one should expect mainly non-regular 3D sTable structures.

The linear $(LiF)_nLi^+$ clusters also show a symmetric atomic charge distribution relative to the central F (n odd) or Li (n even). However, differently from the neutral $(LiF)_nLi$ clusters, the atomic charges on the Li-F units remain practically the same [Li(0.93)F(-0.92)] independently of the distance from the central atom. Besides that, as the chain length increases the charge on the central Li atom or the net charge on the central (Li-F-Li) unit decreases, thus reducing the stabilization due to ion-dipole interactions. The dipole-dipole interactions will also be less effective than in the neutral clusters because in the latter ones the LiF units with larger dipole moments are closer to each other. Hence, relative to the neutral clusters, one should expect a more pronounced decrease in stability as the chain length increases as observed in Figure 11. As for the neutral ones, cyclic planar structures (n > 5) are not sTable and only non-regular planar structures as well as 3D structures can be formed.

Electron affinities (EAs) can be determined for the neutral LiF clusters from the total energies of the $(LiF)_n^-$, $(LiF)_n^0$, and $(LiF)_nF^0$ series following the method described in *Section 3.2.* As expected, these values are equal, in modulus, to the first ionization potential (or detachment energy) of the respective negative ions. Experimental values of electron affinity for molecules and clusters are scarce, but for n = 0, the EA of $(LiF)_nF$ is just, of course, the fluorine EA. By using the laser photo detachment technique, a value of 3.401 eV for the EA of fluorine was obtained, [135] which is in excellent agreement with the theoretical value (3.46 eV) obtained by the method described here. [125]. The dependence of the electron affinity on the cluster size is illustrated in Figure 12 for linear structures, and general trends can be summarized: (i) the members of the $(LiF)_n$ series present lower electron affinities than those calculated for the $(LiF)_nF$ series, and (ii) the electron affinity of both $(LiF)_nF$ and $(LiF)_n$ species increases monotonically with n but levels off around n $\approx$ 5. These results suggest that the neutralization of the $(LiF)_n^-$ ion clusters is much more likely to occur than the neutralization of the $(LiF)_nF^-$ ions and provide an explanation for the fact that the observed desorption yield of the $(LiF)_n^-$ ions is lower than that of the $(LiF)_nF^-$ ions. The electron affinity increases with n because electronic delocalization also increases with the size of the clusters. Such an increase levels off when the delocalization saturates, that is, when the interaction between the two ends of the linear cluster becomes negligible. Note also in Figure 12 that the electron affinity increases by a factor of ~2 as n varies from 0 to 4. Therefore, the process of electron detachment will be more effective for neutralizing the smaller cluster anions formed, thus contributing to a reduction of their desorption yields.

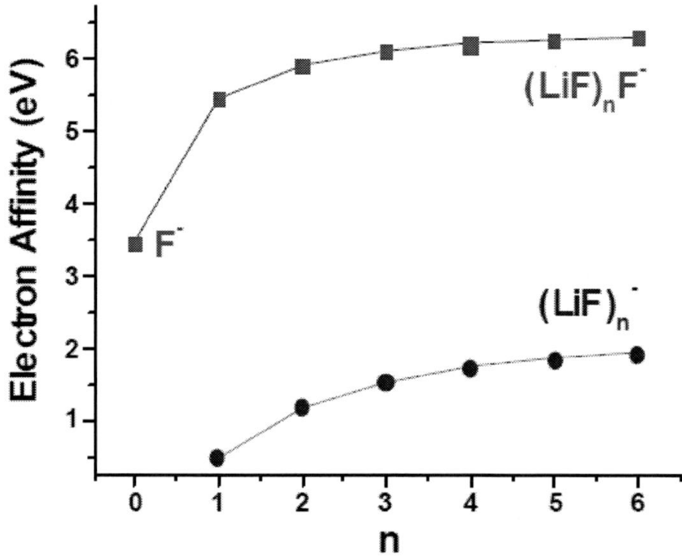

Figure 12. Electron affinity dependence on the cluster size for the $(LiF)_nF^-$ and $(LiF)_n$ species.

The relative stability of the cluster structures can also be analyzed by examining the fragmentation energy as a function of the cluster size. The fragmentation energy, FE, is the difference between the cluster energy and the sum of the energies of the isolated fragments produced by the dissociation of the cluster. Theoretical values of FE have been calculated from the total energies, E_T, for selected fragmentation channels. [122, 125] The dissociation process can be better understood when examined for fragmentation into structures of the same cluster series. In this case, structural rearrangements can be neglected, and the main differences can be attributed to the loss of (XY) units. For example, for linear cluster structures from LiF alkali halide, the fragmentation energy of $(LiF)_nLi^+$, $(LiF)_nF^-$, $(LiF)_nF^0$ and $(LiF)_n^0$ can be defined as:

$$FE_{NT}(n,m) = E_T[(LiF)_mF^+] + E_T[(LiF)_{n-m}^0] - E_T[(LiF)_nF^+] \qquad (4)$$

$$FE_{NT}(n,m) = E_T[(LiF)_mF^-] + E_T[(LiF)_{n-m}^0] - E_T[(LiF)_nF^-] \qquad (5a)$$

$$FE_T(n,m) = E_T[(LiF)_mF^0] + E_T[(LiF)_{n-m}^-] - E_T[(LiF)_nF^-] \qquad (5b)$$

$$FE_{NT}(n,m) = E_T[(LiF)_mF^0] + E_T[(LiF)_{n-m}^0] - E_T[(LiF)_nF^0] \qquad (6)$$

$$FE_{NT}(n,m) = E_T[(LiF)_m^0] + E_T[(LiF)_{n-m}^0] - E_T[(LiF)_n^0] \qquad (7)$$

The $FE_{NT}(n, n - 1)$ values are the binding energies of the precursor cluster $(n - m = 1)$ for LiF^0 emission.

Notice that fragmentation processes may occur with (FE_T) or without (FE_{NT}) charge transfer from the parent to the fragment cluster as illustrated in eq. 5a and b. Figure 13 shows the fragmentation energy dependence on the cluster size for some of the fragmentation channels. In summary, Figure 13 results indicate that (i) large $(LiF)_nLi^+$ and $(LiF)_nF^-$ linear cluster ions decay preferentially by neutral LiF^0 emission; (ii) a successive loss of LiF^0 units is more likely to occur than a single $(LiF)_m^0$ cluster emission; and (iii) fragmentation of $(LiF)_nLi^+$ and $(LiF)_nF^-$ linear cluster ions into Li^0, Li^+, F^0, F^-, $(LiF)Li^+$ or $(LiF)F^-$ is attenuated.

For fragmentation processes that may occur with or without charge transfer, Figure 13 results show that, if no charge transfer occurs during the fragmentation of the $(LiF)_nF^-$ clusters, the most and the least energetically favorable channels are the emission of a neutral LiF unit and of a F^- ion, respectively. If charge transfer from the $(LiF)_nF$ to the $(LiF)_n^-$ fragment occurs, then higher FE values are obtained.

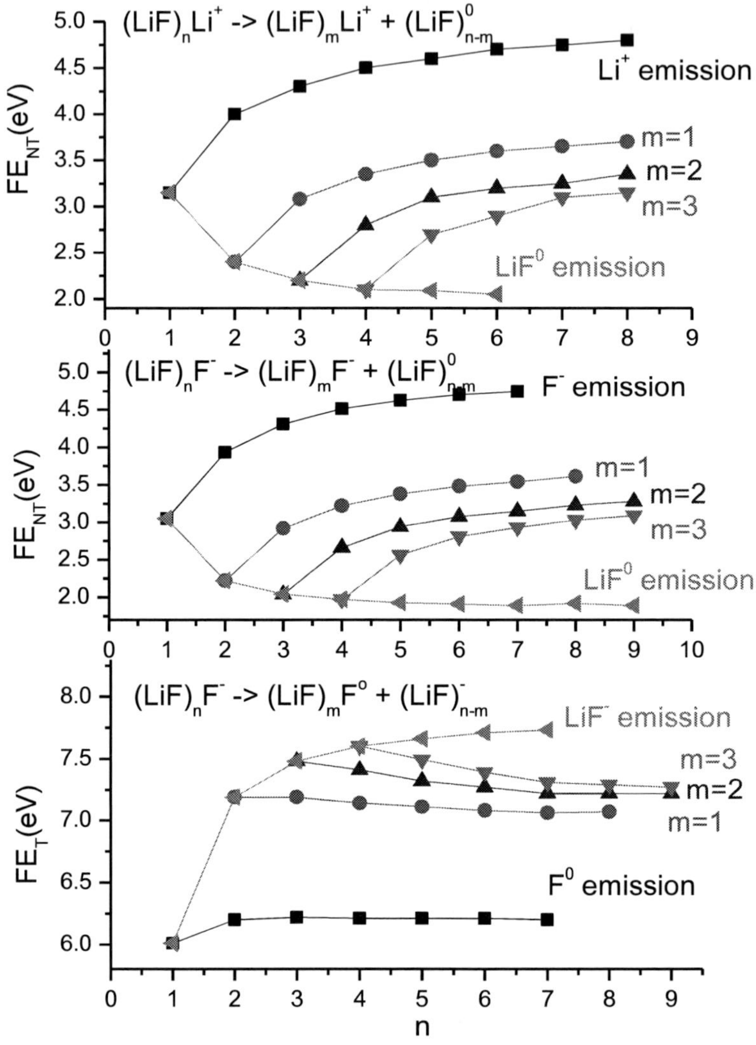

Figure 13. Fragmentation energy dependence on the cluster size for the linear $(LiF)_nLi^+$ and $(LiF)_nF^-$ series.

For the FE_T process, the most energetically favorable channel is the emission of a neutral F^0 atom, whereas the least energetically favorable channel is the emission of a negatively charged LiF^- unit. Comparison of the FE_{NT} and FE_T values suggests that the dissociation without charge transfer is up to 3 times more favorable among the lower energy channels.

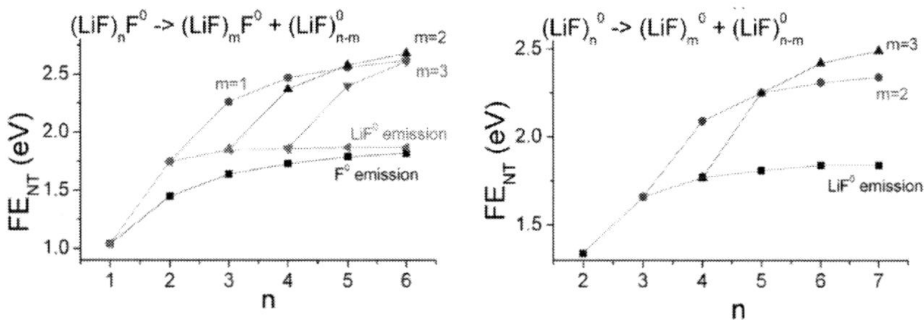

Figure 14. Fragmentation energy dependence on the cluster size for the linear $(LiF)_nF^0$ and $(LiF)_n^0$ series. Left) fragmentation without charge transfer for the $(LiF)_nF^0$ linear series $FE_{NT}(n, m)$. Right): fragmentation without charge transfer for the $(LiF)_n^0$ linear series $FE_{NT}(n, m)$.

Although neither the linear $(LiF)_nF^0$ nor the $(LiF)_n^0$ clusters can be observed directly by mass spectrometry, fragmentation energies, $FE_{NT}(n, m)$, can be predicted using eq. 6 and 7 (see Figure 14). The results suggest that the most energetically favorable channels always involve the smallest possible neutral fragment. For the $(LiF)_nF^0$ series, the preferred emission is the F^0 atom, followed by the emission of a neutral LiF^0 unit. In the case of the $(LiF)_n^0$ series, the smaller the $(LiF)^0$ unit, the smaller the FE value.

The cluster stability can be also discussed using the D-plot methodology for all the series. For example, a D-plot for the $(LiF)_n$ structures is shown in Figure 15. The members of the tetragonal series are sTable for n up to 16, after which the vibration analysis predicts imaginary frequencies. The decrease in stability with the increase of the cluster size for members of this series can be easily observed in the D-plot.

The hexagonal and octagonal series are the most sTable among the tube-like structures investigated and up to $n = 15$ the hexagonal nanotubes are even more sTable than the corresponding cubic and tetragonal members. This same order of relative stability is found at the CCSD/SDD level of calculation indicating that the results obtained at the B3LYP/LACV3P** provide an accurate description of these nanotubes.

Compared to the previous series, the octagon series presents the smallest slope in the D-plot (Figure 15), indicating that the stability of its members decreases much slower with increasing the cluster size when compared to the other series. For example, the member with n = 24 is almost as sTable as the respective member of the hexagon series.

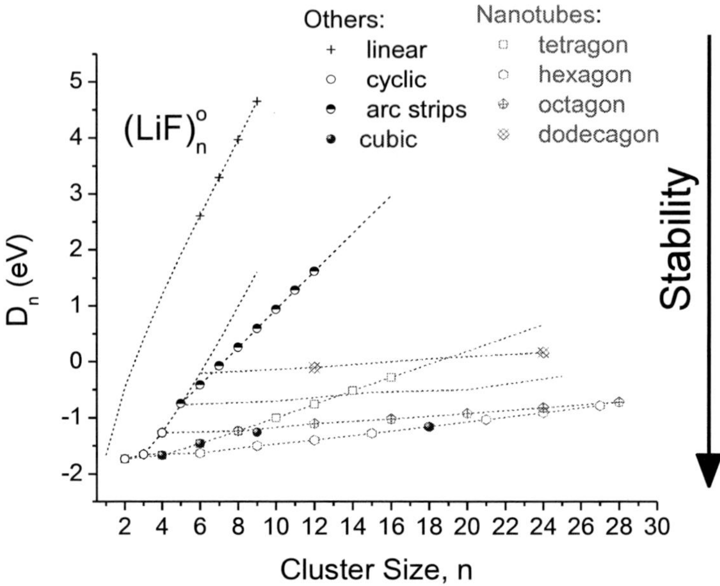

Figure 15. Deviation energy (D$_n$) plot of the (LiF)$_n$ clusters as a function of the cluster size, n. Notice the higher stability (lower D$_n$ values) of the cubic, nanotube/hexagon and nanotube/octagon series. The dotted lines denote extrapolations for the different families, and the symbols correspond to the structures that are real minima.

This suggests that for n > 28, the octagon series may become the most sTable among the analyzed structures. This is indeed the case for n ≥ 32. Preliminary results of quantum dynamics calculations on the octogonal (LiF)$_{28}$ nanotube shows that it is also sTable at room temperature.

The decagon and dodecagon nanotubes are sTable up to *n* = 20 and 24 respectively, after which the vibration analysis predicts imaginary frequencies.

The hexagon and octagon series present the higher stability of all analyzed nanotube series. The addition of new building blocks does not critically compromise the interaction of all the counterparts, creating a good balance between transversal and longitudinal atom interactions. Although one would think that larger macrostructures of ionic crystals need to have symmetrical patterns because of the electrostatic interactions, the boundary conditions of the nanometer scale systems demand that these structures have a unique combination of short- and long- range interactions. Our interpretation of these results is that the nanotube structures provide the "best" symmetry for balancing of periodic building blocks and may be used to create larger macrostructures.

The results obtained for the $(LiF)_n$ clusters show that overall the cubic series is highly sTable as expected from structures which are debris of a crystalline bulk. Surprisingly, however, nanotube clusters with hexagonal and octagonal cross section present similar or higher stability than the corresponding cubic clusters. The stability of cubic structures has been shown to be strongly dependent on the existence of a surface involving a large fraction of the constituents (boundary surfaces), and the distribution of magic numbers (*e.g.,* closed shell in metal clusters) has been attributed mainly to this effect. Nevertheless, in nanotube structures an increase of surface constituents plays a larger role in cluster stability, especially because of their hollow nature. For example, inspection of the lattice parameters of the LiF nanotubes shows that axial stability is smaller than transversal stability on account of a larger parameter size (~ 0.186 nm) when compared to the transversal size (~ 0.178 nm). The axial stability is strongly determined by the amount of stress that the transversal cross section can support making the geometry of the transversal cross section a critical factor.

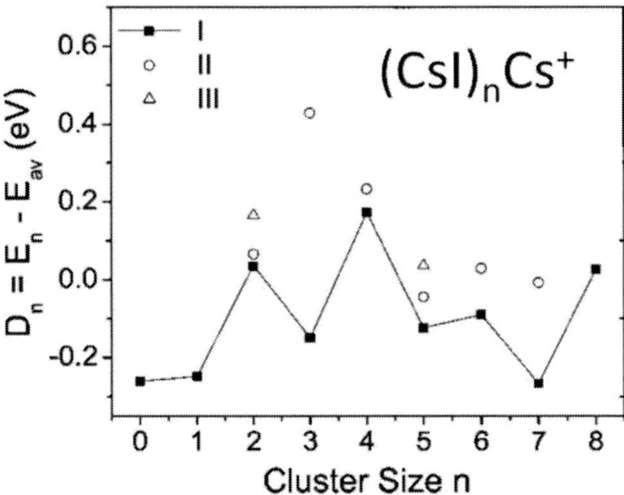

Figure 16. D-Plot: Deviation of the total energy (including the zero point energy correction) as a function of the cluster size, n. (Inset) Total energy as a function of the cluster size, n, is plotted. The connected dots correspond to the lowest energy isomers.

The D-plot for the $(CsI)_nCs^+$ clusters is shown in Figure 16. A relatively high stability is observed for the odd-numbered cluster ions, *e.g.,* n = 3 and n = 7, while cluster ions with n = 2 and n = 4 are less sTable. The higher stabilities of n = 3 and n = 7 have been observed previously in MS experiments. [136]

Inspection of Figures 10 and 16 shows that the cluster structures do not exhibit any bonding pattern that could be related to the stability values, *i.e.,* the charge distribution and bonding distances within the cluster do not present any abrupt change or trend that could be correlated with the stability values. This suggests that the higher stability is a consequence of the intrinsic cluster geometry that makes the electrostatic interaction between its constituents to have a maximum for *n*-odd clusters, *i.e.,* it is a consequence of the convolution of all the short- and long-range interaction between the cluster constituents.

REFERENCES

[1]   R. L. Johnston, Atomic and Molecular Cluster Taylor and Francis, London and New York, 2002.

[2]   G. Stefan, W. Patrick, F. Filip, A. Reinhart, and M. K. Manfred, *J. Chem. Phys.* 116 (2002) 4094.

[3]   P. Gruene, D. M. Rayner, B. Redlich, A. F. G. van der Meer, J. T. Lyon, G. Meijer, and A. Fielicke, *Science* 321 (2008) 674.

[4]   F. Filipp, A. Reinhart, W. Patrick, J. Christoph, G. Stefan, B. Thomas, and M. K. Manfred, *J. Chem. Phys.* 117 (2002) 6982.

[5]   B. K. Vlasta, P. Jiri, B. Marc, and F. Piercarlo, *J. Chem. Phys.* 110 (1999) 3876.

[6]   B. K. Teo, *Polyhedron* 7 (1988) 2317.

[7]   R. A. J. O'Hair and G. N. Khairallah, *J. Cluster Sci.* 15 (2004) 331.

[8]   K. C. Lau and R. Pandey, *Lecture Series on Computer and Computational Sciences* 5 (2006) 116.

[9]   P. Senet, M. Yang and C. *Van Alsenoy, Atoms, Molecules and Clusters in Electric Fields* (2006) 657.

[10] H. C. Chang, C. C. Wu and J. L. Kuo, *Int. Rev. Phys. Chem.* 24 (2005) 553.

[11] T. S. Zwier, *Science* 304 (2004) 1119.

[12] F. A. Fernandez-Lima, C. R. Ponciano, M. A. C. Nascimento, and E. F. D. Silveira, *J. Phys. Chem.* A 110 (2006) 10018.

[13] F. A. Fernandez-Lima, T. M. Cardozo, R. M. Rodriguez, C. R. Ponciano, E. F. D. Silveira, and M. A. C. Nascimento, *J. Phys. Chem. A* 111 (2007) 8302.

[14] H. Gert von, H. Ming-Teh, R. K. Paul, and T. B. Michael, *J. Chem. Phys.* 95 (1991) 3835.

[15] F. A. Fernandez-Lima, C. R. Ponciano, E. F. d. Silveira, and M. A. C. Nascimento, *Chem. Phys. Lett.* 426 (2006) 351.

[16] F. A. Fernandez-Lima, C. R. Ponciano, E. F. D. Silveira, and M. A. C. Nascimento, *Chem. Phys. Lett.* 445 (2007) 147.

[17] C. Lifshitz, *Int. J. Mass Spectrom.* 200 (2000) 423.

[18] K. S. Kim, P. Tarakeshwar and H. M. Lee, Theory and Applications of Computational Chemistry: The First Forty Years (2005) 963.

[19] J. Berkowitz, Photelectron Spectroscopy of Alkali Halide Vapors, Academic, New York, 1979.

[20] S. R. Veličković, J. B. Đustebek, F. M. Veljković, and M. V. Veljković, *J. Mass Spectrom.* 47 (2012) 627.

[21] F. Honda, G. M. Lancaster, Y. Fukuda, and J. W. Rabalais, *J. Chem. Phys.* 69 (1978) 4931.

[22] B. U. R. Sundqvist, P. Håkansson, A. Hedin, D. Fenyö, G. Brinkmalm, P. Roepstorff, and R. E. Johnson, *Methods and Mechanisms for producing Ions from Large Molecules.*, Plenum Press, New York, 1991.

[23] H. Hijazi, L. S. Farenzena, H. Rothard, P. Boduch, P. L. Grande, and E. F. da Silveira, *Eur. Phys. J.* D 63 (2011) 391.

[24] Y. J. Twu, C. W. S. Conover, Y. A. Yang, and L. A. Bloomfield, *Phys. Rev.* B 42 (1990) 5306.

[25] F. A. Fernandez-Lima, C. R. Ponciano and E. F. da Silveira, *J. Mass Spectrom.* 43 (2008) 587.

[26] R. Pflaum and E. Recknagel, *Zeit. für Phys.* D 12 (1989) 249.

[27] G. Wang and R. B. Cole, *Analytica Chimica Acta* 406 (2000) 53.

[28] V. Wysocki, C. Jones, A. Galhena, and A. Blackwell, *J. Am. Soc. Mass Spectrom.* 19 (2008) 903.

[29] F. W. Röllgen and K. H. Ott, *Int. J. Mass Spectrom.* 32 (1980) 363.

[30] D. M. Mann and H. P. Broida, *J. App. Phys.* 44 (1973) 4950.

[31] G. P. Konnen, A. Tip and A. E. de Vries, *Radiat. Eff.* 26 (1975) 23.

[32] F. Honda, G. M. Lancaster and J. W. Rabalais, *Surf. Sci.* 76 (1978) L613.

[33] F. Honda, Y. Fukuda and J. W. Rabalais, *J. Chem. Phys.* 70 (1979) 4834.

[34] X. B. Cox Iii, R. W. Linton and M. M. Bursey, *Int. J. Mass Spectrom.* 55 (1984) 281.

[35] T. M. Barlak, J. E. Campana, R. J. Colton, J. J. DeCorpo, and J. R. Wyatt, *J. Phys. Chem.* 85 (1981) 3840.

[36] T. M. Barlak, J. E. Campana, J. R. Wyatt, and R. J. Colton, *J. Phys. Chem.* 87 (1983) 3441.

[37] I. Katakuse, H. Nakabushi, T. Ichihara, T. Sakurai, T. Matsuo, and H. Matsuda, *Int. J. Mass Spectrom.* 57 (1984) 239.

[38] J. H. Callahan, R. J. Colton and M. M. Ross, *Int. J. Mass Spectrom.* 90 (1989) 9.

[39] S. H. Huh and G. H. Lee, *J. Korean Phys. Soc.* 38 (2001) 107.

[40] L. A. Bloomfield, C. W. S. Conover, Y. A. Yang, Y. J. Twu, and N. G. Phillips, *Zeit. für Phys.* D 20 (1991) 93.

[41] V. M. Collado, C. R. Ponciano, F. A. Fernandez-Lima, and E. F. d. Silveira, *Rev. Sci. Instrum.* 75 (2004) 2163.

[42] F. A. Fernandez-Lima, V. M. Collado, C. R. Ponciano, L. S. Farenzena, E. Pedrero, and E. F. D. Silveira, *App. Surf. Sci.* 217 (2003) 202.

[43] V. M. Collado, F. A. Fernandez-Lima, C. R. Ponciano, M. A. C. Nascimento, L. Velazquez, and E. F. d. Silveira, *Phys. Chem. Chem. Phys.* 7 (2005) 1971.

[44] A. J. Dally and L. A. Bloomfield, *Phys. Rev. Lett.* 90 (2003) 063401.

[45] D. Rayane, A. R. Allouche, I. Compagnon, R. Antoine, M. Aubert-Frécon, M. Broyer, and P. Dugourd, *Chem. Phys. Lett.* 367 (2003) 278.

[46] C. Hao, R. E. March, T. R. Croley, J. C. Smith, and S. P. Rafferty, *J. Mass Spectrom.* 36 (2001) 79.

[47] D. Zhang and R. G. Cooks, *Int. J. Mass Spectrom.* 195–196 (2000) 667.

[48] M. P. Ince, B. A. Perera, and M. J. Van Stipdonk, *Int. J. Mass Spectrom.* 207 (2001) 41.

[49] M. A. Park, E. A. Schweikert, E. F. da Silveira, C. V. B. Leite, and J. M. F. Jeronymo, *Nucl. Instrum. and Meth.* B 56–57, (1991) 361.

[50] M. A. Park, E. A. Schweikert, and E. F. d. Silveira, *J. Chem. Phys.* 96 (1992) 3206.

[51] F. A. Fernandez-Lima, C. R. Ponciano, H. D. F. Filho, E. Pedrero, M. A. Chaer Nascimento, and E. F. da Silveira, *App. Surf. Sci.* 252 (2006) 8171.

[52] C. R. Ponciano, F. E. Ávalos, A. Rentería, and E. F. da Silveira, *Int. J. Mass Spectrom.* 209 (2001) 197.

[53] R. G. Kaercher, E. F. da Silveira, C. V. B. Leite, and E. A. Schweikert, *Nucl. Instrum and Meth.* B 94 (1994) 207.

[54] C. R. Ponciano, R. Martinez and E. F. da Silveira, *J. Mass Spectrom.* 42 (2007) 1300.

[55] R. G. Kaercher, E. F. da Silveira, J. F. Blankenship, and E. A. Schweikert, *Nucl. Instrum and Meth.* B 100 (1995) 383.

[56] R. G. Kaercher, E. F. da Silveira, J. F. Blankenship, and E. A. Schweikert, *Phys. Rev.* B 51 (1995) 7373.

[57] J. P. Thomas, P. E. Filpus-Luyckx, M. Fallavier, and E. A. Schweikert, *Phys. Rev. Lett.* 55 (1985) 103.

[58] E. A. Schweikert, M. G. Blain, M. A. Park, and E. F. Da Silveira, *Nucl. Instrum and Meth.* B 50 (1990) 307.

[59] W. Ens, R. Beavis and K. G. Standing, *Phys. Rev. Lett.* 50 (1983) 27.

[60] B. Schueler, R. Beavis, W. Ens, D. E. Main, and K. G. Standing, *Surf. Sci.* 160 (1985) 571.

[61] J. A. M. Pereira, C. S. C. de Castro, J. M. F. Jeronymo, C. R. Ponciano, E. F. da Silveira, and K. Wien, *Nucl. Instrum and Meth.* B 129 (1997) 21.

[62] O. Almén and G. Bruce, *Nucl. Instrum. and Meth.* 11 (1961) 279.

[63] O. Almén and G. Bruce, *Nucl. Instrum. and Meth.* 11 (1961) 257.

[64] J. E. Campana, T. M. Barlak, R. J. Colton, J. J. DeCorpo, J. R. Wyatt, and B. I. Dunlap, *Phys. Rev. Lett.* 47 (1981) 1046.

[65] M. G. Blain, E. F. da Silveira and E. A. Schweikert, *Journal de Physique* 2 (1989) C2.

[66] M. G. Blain, S. Della-Negra, H. Joret, Y. Le Beyec, and E. A. Schweikert, *Phys. Rev. Lett.* 63 (1989) 1625.

[67] N. Itoh and T. Nakayama, *Nucl. Instrum. and Meth.* B 13 (1986) 550.

[68] N. Itoh, Nucl. *Instrum. and Meth.* B 27 (1987) 155.

[69] G. Betz and K. Wien, *Int. J. Mass Spectrom.* 140 (1994) 1.

[70] J. A. M. Pereira, E. F. da Silveira, and K. Wien, *Rad. Eff. and Def. in Solids* 142 (1997) 247.

[71] J. A. M. Pereira and E. F. da Silveira, *Nucl. Instrum. and Meth.* B 127-128 (1997) 157.

[72] J. A. M. Pereira and E. F. da Silveira, *Nucl. Instrum. and Meth.* B 136-138 (1998) 779.

[73] J. A. M. Pereira and E. F. da Silveira, *Surf. Sci.* 390 (1997) 158.

[74] J. A. M. Pereira, I. S. Bitensky and E. F. da Silveira, *Nucl. Instrum. and Meth.* B 135 (1998) 244.

[75] J. A. M. Pereira, I. S. Bitensky and E. F. da Silveira, *Int. J. Mass Spectrom.* 174 (1998) 179.

[76] J. A. M. Pereira and E. F. da Silveira, *Nucl. Instrum. and Meth.* B 146 (1998) 185.

[77] J. A. M. Pereira and E. F. da Silveira, *Phys. Rev. Lett.* 84 (2000) 5904.

[78] M. Toulemonde, W. Assmann, C. Trautmann, F. Grüner, H. D. Mieskes, H. Kucal, and Z. G. Wang, *Nucl. Instrum. and Meth.* B 212 (2003) 346.

[79] K. Schwartz, C. Trautmann, A. S. El-Said, R. Neumann, M. Toulemonde, and W. Knolle, *Phys. Rev.* B 70 (2004) 184104.

[80] T. Jalowy, R. Neugebauer, K. O. Groeneveld, C. R. Ponciano, L. S. Farenzena, and E. F. da Silveira, *Int. J. Mass Spectrom.* 219 (2002) 343.

[81] T. Jalowy, R. Neugebauer, M. Hattass, J. Fiol, F. Afaneh, J. A. M. Pereira, V. Collado, E. F. da Silveira, H. Schmidt-Bocking, and K. O. Groeneveld, *Nucl. Instrum. and Meth.* B 193 (2002) 762.

[82] T. Jalowy, R. Neugebauer, K. O. Groeneveld, C. R. Ponciano, L. S. Farenzena, and E. F. d. Silveira, *Rev. Sci. Instrum.* 73 (2002) 3187.

[83] T. Jalowy, T. Weber, R. Dorner, L. Farenzena, V. M. Collado, E. F. da Silveira, H. Schmidt-Bocking, and K. O. Groeneveld, *Int. J. Mass Spectrom.* 231 (2004) 51.

[84] H. Hijazi, H. Rothard, P. Boduch, I. Alzaher, F. Ropars, A. Cassimi, J. M. Ramillon, T. Been, B. B. d'Etat, H. Lebius, L. S. Farenzena, and E. F. da Silveira, *Nucl. Instrum. and Meth.* B 269 (2011) 1003.

[85] T. Jalowy, L. S. Farenzena, C. R. Ponciano, H. Schmidt-Bocking, E. F. da Silveira, and K. O. Groeneveld, *Surf. Sci.* 557 (2004) 91.

[86] H. Hijazi, H. Rothard, P. Boduch, I. Alzaher, A. Cassimi, F. Ropars, T. Been, J. Ramillon, H. Lebius, B. Ban-d'Etat, L. Farenzena, and E. da Silveira, *Europ. Phys. J.* D 66 (2012) 1.

[87] W. Assmann, M. Toulemonde, and C. Trautmann, In: *Springer Topics in Applied Physics*, Vol. 110 (R. Behrisch and W. Eckstein, eds.), Springer-Verlag, Berlin, New York, Heidelberg, 2007, p. 401−448.

[88] M. Maier-Borst, P. Loffler, J. Petry, and D. Kreisle, Zeit. *fur Phys.* D 40 (1997) 476.

[89] M. Tabrizchi, *Anal. Chem.* 75 (2003) 3101.

[90] F. A. Fernandez-Lima, C. Becker, K. Gillig, W. K. Russell, M. A. C. Nascimento, and D. H. Russell, *J. Phys. Chem.* A 112 (2008) 11061.

[91] E. W. McDaniel and L. A. Viehland, *Phys. Rep.* 110 (1984) 333.

[92] E. W. McDaniel and E. A. Mason, *Mobility and diffusion of ions in gases,* John Wiley and Sons, Inc., New York, New York, 1973.

[93] A. B. Kanu, P. Dwivedi, M. Tam, L. Matz, and H. H. Hill, *J. Mass Spectrom.* 43 (2008) 1.

[94] K. J. Gillig and D. H. Russell, In: P*atent Cooperation Treaty Int. Appl.* WO0165589, The Texas A and M University System, 2001, p. 36.

[95] K. J. Gillig, B. T. Ruotolo, E. G. Stone, and D. H. Russell, *Int. J. Mass Spectrom.* 239 (2004) 43.

[96] Y. Guo, J. Wang, G. Javahery, B. A. Thomson, and K. W. M. Siu, *Anal. Chem.* 77 (2004) 266.

[97] S. L. Koeniger, S. I. Merenbloom, S. J. Valentine, M. F. Jarrold, H. R. Udseth, R. D. Smith, and D. E. Clemmer, *Anal. Chem.* 78 (2006) 4161.

[98] B. M. Kolakowski and Z. Mester, *Analyst* 132 (2007) 842.

[99] S. D. Pringle, K. Giles, J. L. Wildgoose, J. P. Williams, S. E. Slade, K. Thalassinos, R. H. Bateman, M. T. Bowers, and J. H. Scrivens, *Int. J. Mass Spectrom.* 261 (2007) 1.

[100] F. A. Fernandez-Lima, D. A. Kaplan, J. Suetering, and M. A. Park, *Int. J. Ion Mobility Spectrom.* 14 (2011) 93.

[101] F. A. Fernandez-Lima, D. A. Kaplan, and M. A. Park, *Rev. Sci. Instrum.* 82 (2011) 126106.

[102] P. R. Kemper and M. T. Bowers, *J. Am. Soc. Mass Spectrom.* 1 (1990) 197.

[103] C. Wu, W. F. Siems, G. R. Asbury, and H. H. Hill, *Anal. Chem.* 70 (1998) 4929.

[104] Y. Liu and D. E. Clemmer, *Anal. Chem.* 69 (1997) 2504.

[105] M. F. Jarrold and V. A. Constant, *Phys. Rev. Lett.* 67 (1991) 2994.

[106] J. A. McLean, B. T. Ruotolo, K. J. Gillig, and D. H. Russell, *Int. J. Mass Spectrom.* 240 (2005) 301.

[107] X. Liu, S. J. Valentine, M. D. Plasencia, S. Trimpin, S. Naylor, and D. E. Clemmer, *J. Am. Soc. Mass Spectrom.* 18 (2007) 1249.

[108] C. S. Hoaglund, S. J. Valentine and D. E. Clemmer, *Anal. Chem.* 69 (1997) 4156.

[109] P. Dwivedi, P. Wu, S. Klopsch, G. Puzon, L. Xun, and H. Hill, *Metabolomics* 4 (2008) 63.

[110] F. A. Fernandez-Lima, C. Becker, A. M. McKenna, R. P. Rodgers, A. G. Marshall, and D. H. Russell, *Anal. Chem.* 81 (2009) 9941.

[111] F. A. Fernandez-Lima, H. Wei, Y. Q. Gao, and D. H. Russell, *J. Phys. Chem.* A 113 (2009) 8221.

[112] F. A. Fernandez-Lima, R. C. Blase and D. H. Russell, *Int. J. Mass Spectrom.* 298 (2010) 111.

[113] F. A. Fernandez-Lima, C. Becker, K. J. Gillig, W. K. Russell, S. E. Tichy, and D. H. Russell, *Anal. Chem.* 81 (2009) 618.

[114] C. S. Hoaglund-Hyzer, Y. J. Lee, A. E. Counterman, and D. E. Clemmer, *Anal. Chem.* 74 (2002) 992.

[115] Z. Michalewicz, Genetic Algorithms + Data Structures = Evolution Programs, 1996.

[116] R. S. Zebulum, M. A. C. Pacheco, M. Maria, and B. R. Vellasco, Evolutionary Electronics: Automatic Design of Electronic Circuits and Systems by Genetic Algorithms, CRC Press, Boca Raton, Florida, 2001.

[117] B. Hartke, *J. Phys. Chem.* 97 (1993) 9973.

[118] Y. Xiao and D. E. Williams, *Chem. Phys. Lett.* 215 (1993) 17.

[119] Y. Zeiri, *Phys. Rev.* E 51 (1995) 2769.

[120] D. M. Deaven and K. M. Ho, *Phys. Rev. Lett.* 75 (1995) 288.

[121] A. N. Alexandrova and A. I. Boldyrev, *J. Chem. Theor. Comput.* 1 (2005) 566.

[122] F. A. Fernandez-Lima, O. P. VilelaNeto, A. S. Pimentel, C. R. Ponciano, M. A. C. Pacheco, M. A. C. Nascimento, and E. F. d. Silveira, *J. Phys. Chem.* A 113 (2009) 1813.

[123] J. D. DeBord, S. Della-Negra, F. A. Fernandez-Lima, S. V. Verkhoturov, and E. A. Schweikert, *The J. Phys. Chem.* C 116 (2012) 8138.

[124] F. A. Fernandez-Lima, M. A. C. Nascimento and E. F. da Silveira, *Nucl. Instrum. and Meth.* B, 273 (2012) 102–104.

[125] F. A. Fernandez-Lima, O. P. Vilela Neto, A. Silva Pimentel, M. A. C. Pacheco, C. R. Ponciano, M. A. Chaer Nascimento, and E. F. da Silveira, *J. Phys. Chem.* A 113 (2009) 15031.

[126] F. A. Fernandez-Lima, C. Becker, K. Gillig, W. K. Russell, M. A. C. Nascimento, and D. H. Russell, *J. Phys. Chem.* A 112 (2008) 11061.

[127] N. Rosch and S. B. Trickey, *J. Chem. Phys.* 106 (1997) 8940.

[128] C. M. Breneman and K. B. Wiberg, *J. Comp. Chem.* 11 (1990) 361.

[129] Schroedinger Inc., Portland, OR, 2004.

[130] GAMESS/US, *Comp. Phys. Commun.* 149 (2002) 21.

[131] F. A. Fernandez-Lima, A. V. Henkes, E. F. da Silveira, and M. A. C. Nascimento, *J. Phys. Chem.* C 116 (2012) 4965.

[132] A. Aguado, A. Ayuela, J. M. López, and J. A. Alonso, *Phys. Rev.* B 58 (1998) 9972.

[133] F. A. Fernandez-Lima, C. R. Ponciano, G. S. Faraudo, M. Grivet, E. F. da Silveira, and M. A. C. Nascimento, *Chem. Phys.* 340 (2007) 127.

[134] F. A. Fernandez-Lima, C. R. Ponciano, E. F. da Silveira, and M. A. C. Nascimento, *Chem. Phys. Lett.* 426 (2006) 351.

[135] C. Blondel, P. Caccioni, C. Delsart, and R. Tranhauer, *Phys. Rev.* A 40 (1989) 3678.

[136] R. Pflaum, K. Sattler and E. Recknagel, *Phys. Rev.* B 33 (1986) 1522.

In: Halides
Editor: Jean Lefebure

ISBN: 978-1-62417-947-1
© 2013 Nova Science Publishers, Inc.

Chapter 2

SYNTHESIS, CHARACTERIZATION AND INFLUENCE OF ELECTROLYTE SOLUTIONS TOWARDS THE ELECTRICAL PROPERTIES OF NYLON-6,6 NICKEL CARBONATE MEMBRANE: TEST FOR THE THEORY OF UNI-IONIC POTENTIAL BASED ON THERMODYNAMICS OF IRREVERSIBLE PROCESSES

Tanvir Arfin[1], and Faruq Mohammad[2,3]*

[1]PGM Group, Chemical Resource Beneficiation (CRB) Research
Focus Area, North-West University, South Africa
[2]Department of Environmental Toxicology,
Southern University and A&M College, Baton Rouge, Louisiana, US
[3]Center for Advanced Microstructures and Devices,
Louisiana State University, Baton Rouge, Louisiana, US

*Corresponding Author. Tel.: +27 (0) 18 299 1576; Fax: +27 (0) 18 299 1667.Email address:
tanvirarfin@ymail.com (T. Arfin), faruq_mohammad@subr.edu (F. Mohammad).

ABSTRACT

In this study, the nano-crystalline organic-inorganic composite nylon-6,6 nickel carbonate membrane was synthesised,subsequent to which (a) the physico-chemical characteristics of the membrane was evaluatedby employing ATR-FTIR, SEM, EDX, TEM,TGA, XRD, and porosity measurements, and (b) membrane potential measurements were carried out using different concentrations of KCl, NaCl and LiCl 1:1 electrolyte solutions($2.5 \times 10^{-2} \leq c$ (M)$\leq 1 \times 10^{0}$).For the potential measurements, the high C_2 concentrations of the electrolyte were maintained on one side of the membrane and low C_1 concentration of the same electrolyte on the other side of the membrane, where the solution-concentration ratio $\left(C_2/C_1\right)$ was maintained at 10 and the membrane potentials found to be higher at low pH values. Also, we observed the positive membrane potentials for the composite membrane and is an indication of the membrane to be negatively charged and this potential found to be dependent of cation selectivity and is decreasing in the order LiCl>NaCl>KCl. At low electrolyte concentrations, the membrane potential found to be high and with increase of concentration, the membrane potential deceased. In addition to the potential measurements, the successful application of Teorell-Meyer and Sievers (TMS), Kobatake et al. and Nagasawa et al. theories were also employed to estimate the charge density of the membrane immersed in different electrolytes and it was found that the charge densities decreased in the order KCl>NaCl>LiCl for 1:1 electrolytes. The extended TMS theory was used to investigate the transference numbers, mobility ratio, distribution coefficients, charge effectiveness, perm selectivity and concentration of fixed charge for 1:1 electrolyte solutions.

1. INTRODUCTION

Today the field of nanotechnology is rapidly sweeping through all vital fields of science and thechnology. At the nanoscale, the physical and chemical properties differ from the properties of individual atoms and molecules of bulk matter and therefore, nanotechnology provides opportunities to develop new classes of advanced materials. The composite nano materials plays an important role in many environmental and industrial applications, as it consists of two or more physically distinct components and exhibits properties that are entirely different from their individual component. These materials possess fully controlled functionality and hydrophobicity which can be used to open new avenues for the organo-metallic chemistry [1].

Nylon-6,6 was first electrochemically synthesized more than two decades ago [2] and has attracted great interst due to its environmental stability to oxygen and water, ease of synthesis, and high conductivity [3]. The potential applications of Nylon-6,6 are in the preparation of actuators, chemical and biosensors, electrodes, and electronic devices[4-8]. Similarly, Nickel carbonates were used in industry as catalysts and can occur as minerals in various natural deposits[9,10].

The membrane transport properties are determined by the inner membrane morphology. For this reason, the understanding of relationships between the behavior of membranes and their structures is fundamental, and numerous works have been carried out to study the relationship between membrane properties and structure in different media [11,12]. The measurement of the membrane potential is a significant method for characterizing the ion transport phenomena through a charged membrane. The membrane potential for an electrolyte is very important in terms of both theoretical and practical significance and it depends on the concentration, temperature. The membrane charge interacts with ions electrostatically and affects the separation efficiency of ions through the partition of ions from the bulk solution into the membrane. The electrical properties of the membrane around the interface are, therefore, very important for the separation efficiency of a membrane.The characterization of membrane charge has been investigated by several researchers theoretically [13,14] and experimentally [15,16].

In this article, we performed the synthesis and characterization of nylon-6,6 nickel carbonate membranes via a sol–gel approach by using nylon-6,6 as a binder. The charged state of the membrane is well reflected in the membrane potential, since the membrane potential is determined by the partition of ions into the membrane as well as the mobilities of ions within the membrane.The observed membrane values were found to be inversely proportional to the electrolyte concentrations and increased with decrease in the external electrolyte concentration. This indicates that the membranes are cation selective. The charge density was most effective parameter that controlled the membrane phenomena and was calculated using the membrane potentials values for different electrolytes by using different theories such as Teorell, Meyer, and Sievers (TMS), Kobatake and Nagasawa. In addition, other parameters such as the transport numbers, mobility ratio, and distribution coefficient etc were also calculated.

2. EXPERIMENTAL

2.1. Reagents and Instruments

The reagents for the synthesis of nylon-6,6 nickel carbonate were obtained from Sigma-Aldrich and Merck Chemicals (South Africa Ltd.) and all the chemicals were used as received. All other reagents and chemicals were of analytical grade and were used without any further purification.

A refrigerated thermostat (Julabo F12-ED), hydraulic press (Carver Hydraulic Unit Model 3912, Wabash, USA), pH meter (744 pH Meter), magnetic stirrer (MSH300, Boeco, Germany), digital multimeter (T235H DIY Digital Multimeter, HellermannTyton), reference electrodes (Radiometer analytical), vacuum oven (Binder VD25), milli-Q water (Millipore, Inc.), ball mill (Crescent, Wig-L-Bug), Fourier-transform infrared (FT-IR) spectrometer (Bruker Alpha ATR), X-ray powder diffraction (Philips , PW 3040/60 X'Pert Pro), FEI QUANTA 200 ESEM with an Integrated OXFORD X-SIGHT EDS System, mercury intrusion porosimetry (Micromeritics, Model AutoPore IV 9500 V), and transmission electron microscopy (Philips CM10 electron microscope) were employed.

2.2. Synthesis of Nylon-6,6Nickel Carbonate Membranes

Nickel carbonate precipitates were prepared by mixing a 0.2 M nickel (II) chloride (99.98%) solution and a 0.2 M sodium carbonate (99.98%) solution yielding light greenish precipitate. The pH of the mixture was adjusted to 1.0 by adding dilute hydrochloric acid (32%) under constant stirring for 24h and the resultant precipitate was kept for another 24h at room temperature (25±0.2°C) for digestion. The supernatant liquid was decanted and the precipitates were filtered by suction. The excess acid was removed by several washings with milli-Q water and the materials were dried in vacuum oven over P_4O_{10} at 40°C. A fine powder was obtained using a pestle and mortar after which the resultant powder was sieved using a 0.0029 inches sieve.

Nylon-6,6 (99.98%) precipitates were prepared in an acidic medium [17]. The resultant powders of nylon-6,6 were also ground and sieved through a 0.0029 inches sieve. Different proportions of nylon-6,6 and nickel carbonate powder were mixed thoroughly using a grinder as indicated in Table 1. The fine composite powder was finally obtained by grinding the material with a ball mill for 30 mins. The mixture was placed in an oven at 125±0.1°C for about 30 min to equilibrate the reaction mixture [18]. All the membranes were prepared by pressing the composite into pellets employing a hydraulic press.

2.3. Experimental Setups

The membrane potential measurements were carried out in a test cell similar to that described elsewhere [19]. The established membrane potentials of the cell are: Hg/HgO|MCl (C_1)| Membrane|MCl (C_2)| Hg/HgO and the membrane potentials of this cell were measured using the borax cell. The membrane area was 196.7 mm^2 and the volume of each half cell was about 25 ml. The membranes were tightly clamped between the two half-cells, which were filled with the solution in the different electrolye concentration $(2.5 \times 10^{-2} \leq c\ (M) \leq 1 \times 10^{0})$. The solution-concentration ratio (C_2/C_1) was maintained at 10, where C_1 and C_2 were the electrolyte concentrations in bulk solutions 1 and 2, respectively. Experiments were carried out using MCl (M = K, Na and Li) aqueous solutions at different concentrations and constant pH (pH = 6.8±0.2). All electrolytes were of analytical grade and solutions were prepared with milli-Q Quality water. In order to minimize the concentration-polarization at the membrane surfaces, a magnetic stirrer was placed at the bottom of each half-cell, which allows an external control of its speed rate. Measurements were carried out at a stirring rate of 500 rpm under isothermal conditions at 25±1°C. The Hg/HgO electrode connected to the digital multimeter was used to measure the membrane potential.

2.4. Instrumental Characterization

The surface of the synthesized membrane was observed with scanning electron microscopy (SEM). For SEM, the sample was dispersed onto a carbon tape and thereafter platinum coated using a sputter coater system to prevent charge accumulation onto the sample.

The chemical composition of the synthesized membrane was determined by energy-dispersive X-ray spectroscopy (EDS) with INCA Energy system being installed in the SEM. The samples were deposited on a standard aluminium holder.

The fourier-transformation infrared (FT-IR) spectrum of sample was recorded by ATR technique with Bruker Alpha spectrometer in the range of 400-4000 cm^{-1} at the scan rate of 32 scans/min witha resolution of 4 cm^{-1}.

The crystallographic characterization of synthesized membrane was carried out with a powdered X-ray diffraction (XRD) using Cu-Kα radiation of wavelength 1.5405 Å with 40 kV, 45 mA and step size of 0.1 per 3s over the range of $2\theta = 5–80°$.

The particle size of the membrane was investigated with a transmission electron microscopy (TEM). Sample was prepared by placing a drop of working solution on a carbon-coated standard copper grid operating at 100 kV. The size of the particles was calculated using the scale provided in the micrograph.

Pore size of the synthesized membrane was determined using the Porosimeter.

3. RESULTS AND DISCUSSION

Composite nylon-6,6 nickel carbonate membrane was prepared by mixing inorganic precipitate of nickel carbonate and nylon-6,6 in different proportions as given in Table 1. S-5 was selected for detailed studies on the basis of its high potential, reproducible behaviour and mechanical stability compared to the other samples prepared.

From the above, it is very clear that membrane potentials depend upon the relative proportion of the binder (nylon-6,6). Membranes having higher binder content possess good mechanical strength but lack in their electrochemical properties; whereas the membranes having lower binder content possess good electrochemical properties, but their mechanical properties are comparatively poor. Both these mechanical and electrochemical properties also depend on the particle size of the binder. So to obtain a better membrane, a compromise between the electrochemical and mechanical properties is needed, i.e. an optimization of the binder ratio and particle size is therefore very essential [20,21]. Thus in our analysis, the membrane consisting of 25% nylon-6,6 was found to show the best mechanical stability as well as electro-chemical performance.

The material found to be sTable in some of the organic solvents (DMSO, DMF, CH$_3$CN, m-Cresol, THF), acids (HCl, HNO$_3$, HClO$_4$ and H$_2$SO$_4$) and bases (NaOH, KOH) up to 1 M concentration, while it was found to be miscible in CCl$_4$.

Table 1. Preparation conditions and the membrane potentials for various samples of nylon-6,6 nickel carbonate membrane

Samples	Mixing volume ratio		Membrane Potential		
	0.2 mol dm^{-3} NiCO$_3$ (g)	Nylon 6,6 (g)	KCl (mV)	NaCl (mV)	LiCl (mV)
S-1	0.147	0.103	5.32		
S-2	0.157	0.093	5.51		
S-3	0.167	0.083	5.64		
S-4	0.177	0.073	5.68		
S-5	**0.187**	**0.063**	**5.74 ± 0.01**	**5.82 ± 0.015**	**5.95 ± 0.011**
S-6	0.197	0.053	5.66		
S-7	0.207	0.043	5.62		

Bold signifies the selected sample (on the basis of membrane potential with standard deviation) employed in further experimental studies mentioned in this report.

Figure 1. SEM of nylon-6,6 nickel carbonate membrane material.

The SEM is a versatile imaging technique capable of producing three-dimensional profiles of the material's surface. SEM is used in this study to exact the quantititaive and qualitative information pertaining to agglomerate size/shape, particle morphology. Since the nanomaterials are thermo-dynamically unsTable and have a tendency to form agglomeration that leads to the attachment of particles at edges or corners. It is also decribed as a number-reducing process which promotes mass-conservation and which increases the particle size distribution. So, the agglomeration decreases the particle surface area available for chemical reaction. In our case, the SEM of nylon-6,6 nickel carbonate surface was performed after coating with platinum in vacuum for 20 min and is shown in Figure 1. It is clear from the Figure that the irregular aggregation of the particle is observed because the SEM sample was prepared with an isolated dry powder. However, it could be seen that sample nanocrystals are almost spherical to form smaller and larger particles, and this makes the size estimation to be difficult.

$$NiCO_3 \cdot nH_2O + (-CO-(CH_2)_4-CO-NH-(CH_2)_6-NH-)$$

$$\downarrow$$

$$[NiCO_3 \cdot (-CO-(CH_2)_4-CO-NH-(CH_2)_6-NH-)] \cdot nH_2O$$

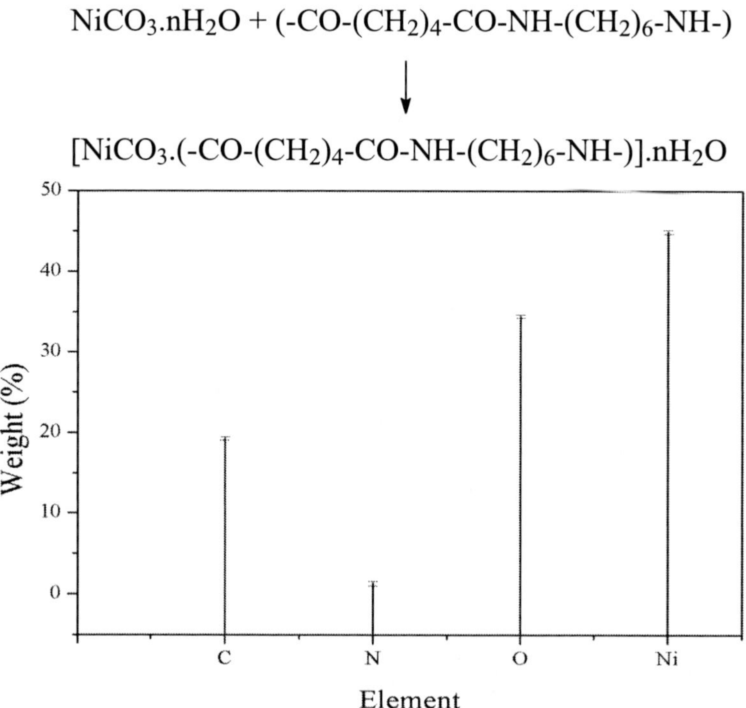

Figure 2. EDS of nylon-6,6 nickel carbonate membrane material. Data represents the mean ± SD for three individual experiments.

The backscatter electron images in the SEM display compositional contrast that results from different atomic number elements and their distribution. EDS allows one to identify what those particular elements are and their relative proportions. Chemical microanalysis was carried out on different parts of the material by Environmental scanning electron microscope with energy-dispersive spectroscopy (ESEM-EDS). The elemental composition of membrane material obtained from this is shown in Figure 2. Three randomly selected spots on the samples were averaged. The standard deviation and the mean values show that Ni, O, N, and C are quite uniformly distributed in the sample. It reveals the high and low proportion of Ni and N on the membrane and based on the results, the following tentative structure is proposed:

The IR spectrum of the material gives more information than is normally available from the electronic spectra. It is used for the rapid detection of functional groups. In this technique, almost all the groups absorbs characteristically within a definite range. Identical materials have identical IR spectra. Molecules with identical or similar shapes of their IR spectra in the finger print region have the same or a similar skeleton of atoms. The shift in the position of absorption for a particular group may change with the change in the structure of the molecules. Thus conjugation of multiple bonds shifts the IR absorption to lower wave numbers.

The FT-IR spectrum of the nylon-6,6 nickel carbonate membrane material is shown in Figure 3. From the Figure, the intense and adsorption band centred at 3638 cm^{-1} corresponds to the stretching vibration of the hydroxyl group. A band close to 1582 cm^{-1} corresponds to the deformation mode (δH_2O) of water molecules. Generally, strong infrared modes around 1071 cm^{-1} are due to the symmetric stretching vibration (v_1), of the carbonate groups, while intense infrared peaks near 1385 cm^{-1} are due to the antisymmetric stretching (v_3). Infrared modes near 801 cm^{-1} are caused by the out-of-plane bend (v_2). Infrared modes near 683 cm^{-1} are due to the in-plane bending mode (v_4).This mode is doubly degenerate for undistorted CO_3^{2-} groups [22-24]. As the carbonate groups become distorted from regular planar symmetry, thus, infrared spectroscopy provides a sensitive test for structure distortion of CO_3^{2-} groups. The bands at 2932 (asymmetric methylene stretching mode: v_{as} (CH_2)) and 2859 cm^{-1} (symmetric methylene stretching mode: v_s (CH_2)) are not related to crystalline fraction [25]. These bands has discussed in terms of the conformational change and lateral chain-chain interaction [26].The α-phase shows characteristic absorbance peaks at ca. 1600 (amide I: (C=O)), 1582 (amide II: coupling of C-N stretching and N-H in plane bending), 1230 cm^{-1}

(amide III) and 1198 cm^{-1} as seen in the region from 900 to 1800 cm^{-1}. The bands at 1362 and 934 cm^{-1} are CH_2 wagging vibration and intrinsic band to CH_2CO group, respectively. The amide V ((N-H) wagging mode; w (N-H)) and amide VI ((C=O) wagging mode; w (C=O)) bands appear at 683 and 579 cm^{-1} respectively, for the characteristic of α-phase. The band at 725 cm^{-1} corresponds to the rocking mode of CH_2. Both α- and γ-phase show this band at the same wave number, indicating the same conformation of CH_2 segments when the different hydrogen bonding schemes are taken place [27]. An assembly of small peaks around 831cm^{-1} corresponds to the superposition of metal–oxygen stretching vibrations for nylon 6,6 [28].

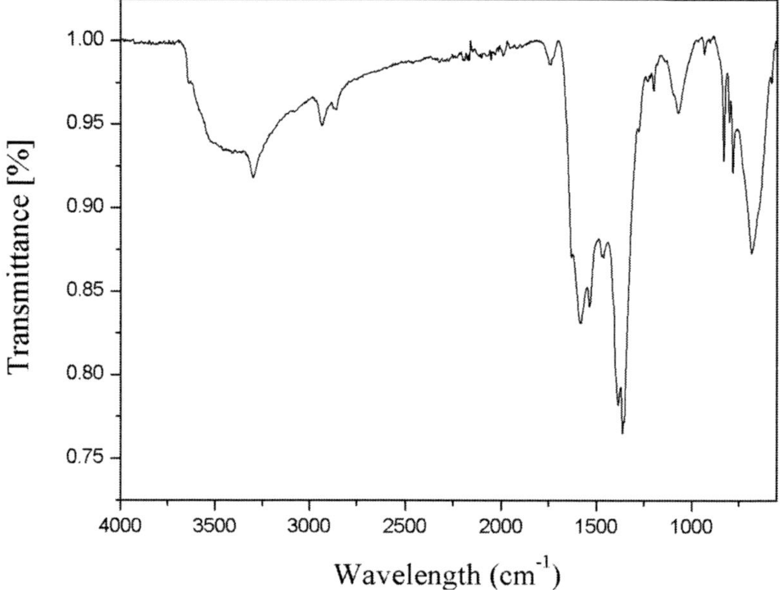

Figure 3. FT-IR spectrum of nylon-6,6 nickel carbonate membrane material.

XRD is a powerful tool for the study of the crystallinity and atomic structure of materials and forms an integral part in a comprehensive characterization study of nanophase material. It is extensively used for the determination of Bravais lattice types and unit cell dimensions. The dimensions of x-ray wavelengths are in the same order as nanoparticle sizes. For the purpose of this study, the XRD tools were used in the investigation of crystal structure and size. The average crystalline size was determined on the basis of the well-known Sherrer equation.

$$D = \frac{0.9\lambda}{\beta\cos\theta} \tag{1}$$

where 0.9 is the shape factor, λ the X-ray wavelength, θ is the Bragg angle and β the full-width of the diffraction line at one half of the maximum intensity.

The powder XRD pattern of the nylon-6,6 nickel carbonate membrane material is shown in Figure 4. From the Figure, the XRD pattern of the material exhibits several strong peaks at 2θ values of 15.52°, 18.07°, 20.40°, 24.12°, 26.14°, 32.64° and so on. The observation of strong peak intensities shows that the as prepared material is crystalline in nature and the average crystalline size of the material lies around 91.5 nm.

TEM probes the internal structure of solids and gives access to micro-structure in detail. It is almost exclusively used in the investigation of the average particle size, particle shape, and particle size distribution of nano, micro sized materials. Figure 5 shows the TEM image of the nylon-6,6 nickel carbonate membrane material and from the image, the membrane polymer appears to be dark (black) due to the scattering off electrons taking place through the polymer, while the pores appear to be bright (white) due to the lack of such scattering. The particles were irregular in shape, agglomeration, and the size falling in the range of nanometers (from scale of 100nm).

Porosity is a fraction of the total volume of a material that is occupied by pores and voids. The pore volume can be described as the total internal volume per unit mass of the material. It is important to determine the pore size and its distribution. The porosity of the support influences the diffusion of reactants and products on the surface [29]. The pore size distribution recorded using nylon-6,6 nickel carbonate (Figure 6) shows only a single broader peak, macro porous with an average pore size centred at 226.47 nm and total porosity was found to be 60.02%. This erroneous average pore size illustrated the inconsistences in the determination of the average pore size using Porosimetery.

When an ionic gradient is maintained using two solutions of different concentrations of the same electrolyte on either side of the membrane, the diffusion of electrolytes from the region of high concentration to low concentraion takes place. There will be flow of water in the opposite direction. Further, an electric field due to differences in the ionic mobilities is established through the membrane. If the membrane contains a large number

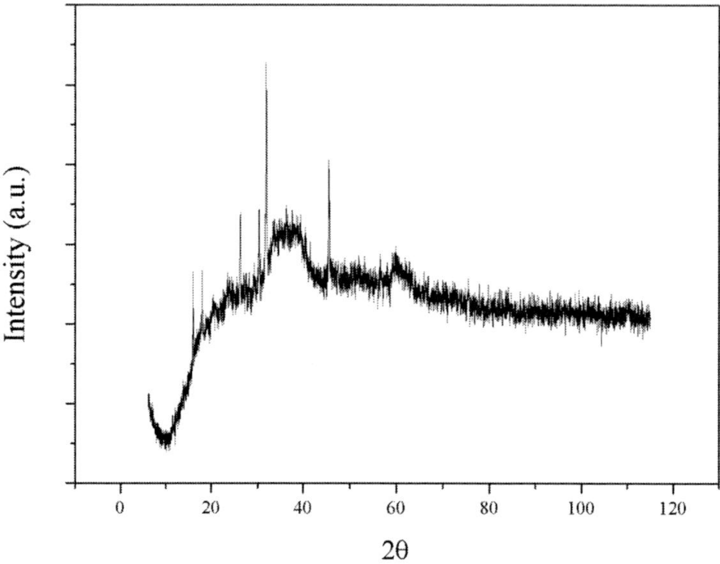

Figure 4. XRD of nylon-6,6 nickel carbonate membrane material.

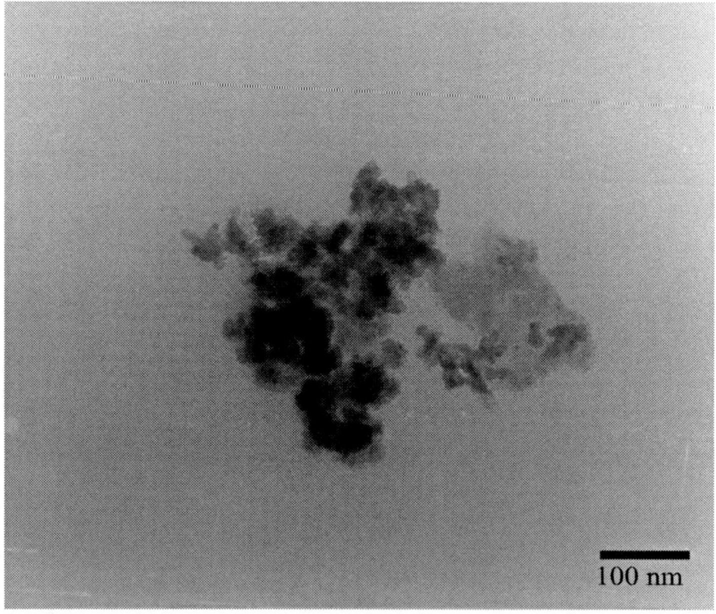

Figure 5. TEM image of nylon-6,6 nickel carbonate membrane material.

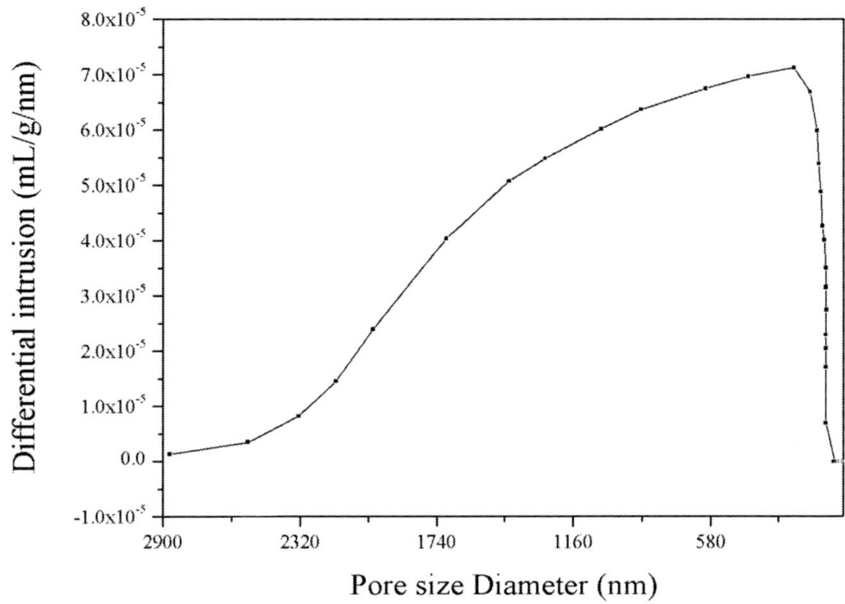

Figure 6.Pore size distribution of nylon-6,6 nickel carbonate membrane material.

of fixed groups, say negatively charged, the potential gradient (dilute solution side positive) will be many times larger than the liquid junction potential, which is normally observed when the two similar solutions are carried together the charge or uncharge membrane in between. These transport phenomena are often described by some extended forms of Nernst-Flux equations [30].

The concentration potentials, i.e. potentials arising across the layer-type membranes when different concentrations of various 1:1 electrolyte solution are placed at $25\pm1°C$, are shown in Table 2.

Generally, the membrane has the unique ability to generate potentials when interpose between two electrolyte solutions of different concentrations [19,31] due to the presence of a net charge on the membrane. The generation of such potentials is always dependent on the porosity of the membrane. If the membrane pores are too wide to produce good potentials due to large amount of charge on the membrane, then the pores may not be suiTable to produce ideal potentials.

Table 2. The observed values of membrane potentials for nylon-6,6 nickel carbonate membrane using 1:1 electrolytes at 25±1°C

Electrolyte Concentration C_2 (mol dm^{-3})	Membrane potential (mV)		
	KCl	**NaCl**	**LiCl**
1.0	5.74 ± 0.01	5.82 ± 0.015	5.95 ± 0.011
0.7	5.78 ± 0.015	5.95 ± 0.015	6.05 ± 0.01
0.5	5.79 ± 0.015	6.03 ± 0.015	6.14 ± 0.015
0.25	5.81 ± 0.011	6.16 ± 0.015	6.24 ± 0.005
0.10	5.88 ± 0.012	6.24 ± 0.02	6.33 ± 0.017
0.07	5.92 ± 0.015	6.30 ± 0.01	6.45 ± 0.02
0.05	5.99 ± 0.01	6.39 ± 0.01	6.58 ± 0.005
0.025	6.28 ± 0.005	6.46 ± 0.012	6.67 ± 0.015

Each experiment was repeated three times to ensure the accuracy of measurements and the data are shown as mean ± SD's of three individual experiments.

In this case, the pore diameters of the nylon-6,6 nickel carbonate membrane was found to be approximately 127.2 nm, and this value was derived from the porosity measurement. Observed membrane potentials are low (+ve). The membrane potential decreases with an increase in electrolyte concentrations and depends on the electrolyte species or the mobilities of ions involved. Thus, the membrane is negatively charged and selectivity increases with dilution due to structural changes produced in the electrical double layer at the solution membrane interface. Potential values (mV) for the electrolytes follow the sequence for the cations [32] in the order $Li^+ > Na^+ > K^+$ and is in accordance with the hydrated radii of ion. This is attributed to increasing the speed of ion in the pore as the size of pores decseses with increasing potential. The magnitude of the Nersnst potential is determined by the ratio of the concentrations of that specific ion on the two sides of the membrane. The greater this ratio, greater will be the tendency for that ion to diffuse in one direction and therefore, greater the Nernst potential required to prevent the diffusion. Thus, the ionic atmosphere of fixed charge is negligible at high concentration, the Cl$^-$ ions might pass through the pores easily.

At the interface between membrane and electrolyte solutions, the Donnan potential occurs due to the transfer of ions. Inside the membrane, the diffusion potential arises since ions would diffuse from the high concentration side to the low concentration under a certain concentration gradient. Membrane potential is the summation of the Donnan potential and the diffusion potential, and it can also be named as the exclusion-diffusion potential [33]. Membrane potential can be measured directly by determining the electrical properties of a membrane or the activities of ions present in the membrane.

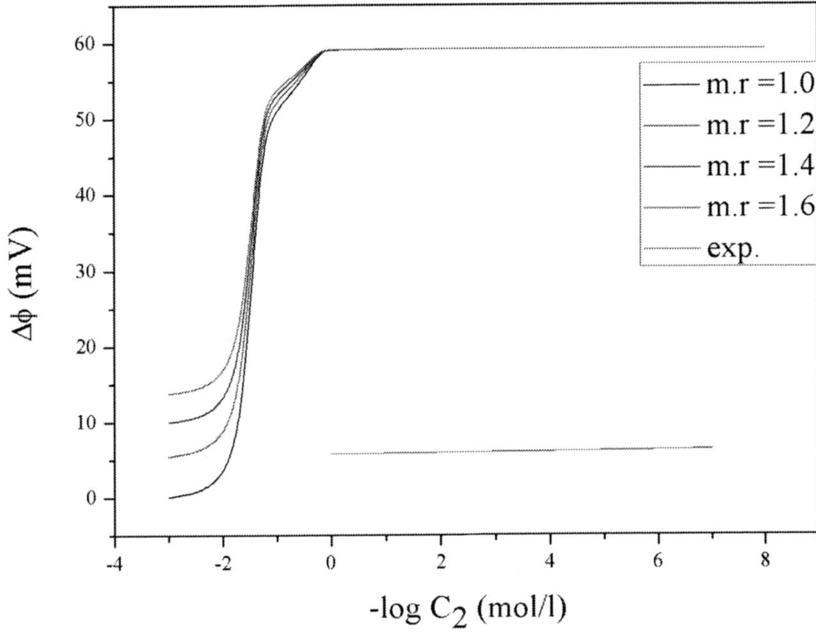

Figure 7. Plots of membrane potentials versus –log C2 for the nylon-6,6 nickel carbonate membrane using KCl electrolytes. Smooth curves are the theoretical concentration potentials (X = 1), with m.r. referring to the mobility ratio.

Table 3. Derived values for nylon-6,6 nickel carbonate membrane parameters of charge density using TMS, Kobatake and Nagasawa theory

Electrolyte	TMS		Kobatake		Nagasawa
	$\left(\dfrac{\bar{u}}{\bar{v}}\right)$	X (eq/l)	χ_c (eq/l)	χ_d (eq/l)	θ (eq/l)
KCl	1.20	0.62	0.53	0.48	0.57
NaCl	1.20	0.36	0.39	0.32	0.45
LiCl	1.20	0.34	0.32	0.27	0.39

Teorell [34,35] -Meyer and Sievers (TMS) [36-38] considered the membrane as a charged barrier. The important features of TMS theory have been nicely described by Lakshminarayanaiah [39] which are used by various investigators including Beg and coworkers in a number of studies with an ion exchange membranes[40,41]. The TMS theory assumes the following: (a) the electrolytes dissolve perfectly, (b) the activity coefficient of all ions is equal to unity, (c) the electrical charge of the membrane is uniform, (d) the osmotic pressure of the hydrostatic flow of water is negligible, (e) the effect of the diffusion boundary layer at membrane-solution interfaces is negligible, (f) the charge groups are homogeneously distributed in the membrane, (g) the cation, anion mobilities and fixed charge concentration are constant throughout the membrane phase (h) the cation, anion mobilities and fixed charge concentration are independent of the salt concentration.

According to this theory, the membrane potential is considered to be composed of two Donnan potentials at two solution membrane interfaces and a diffusion potential arising from the unequal concentrations of the mobile ions at the two membrane phases. For a highly idealized system, these authors have derived the following equation for membrane potential at 25±0.1°C:

$$\Delta\phi = 59.2\left(\log\frac{C_2}{C_1}\frac{\sqrt{4C_1^2 + X^2} + X}{\sqrt{4C_2^2 + X^2} + X} + \overline{U}\log\frac{\sqrt{4C_2^2 + X^2} + X\overline{U}}{\sqrt{4C_1^2 + X^2} + X\overline{U}} \right) \quad (1)$$

$$\overline{U} = \left(\frac{\overline{u} - \overline{v}}{\overline{u} + \overline{v}} \right) \quad\quad\quad (2)$$

where $\overline{u}$ and $\overline{v}$ are the cation and anion mobilities (m^{2}v^{-1}s^{-1}) , respectively in the membrane phase. C_1 and C_2 are the concentrations of the solutions on either side of the membrane and X is the charge density expressed in equivalents per unit volume of membrane. This unit (X) finds its origin in the electro-neutrality requirements of the membrane phase.

In order to evaluate this parameter for a simple case of 1:1 electrolyte and a membrane carrying net negative charge of unity ($X = 1$), theoretical concentration potentials $\Delta\phi$ existing through the membrane were calculated as a function of C_2, the ratio C_2/C_1 being kept at a constant value of 10 for different mobility ratios $\bar{u}/\bar{v}$ and plotted as shown in Figure 7. The observed membrane potential values with various 1:1 electrolyte solutions were also plotted on the same graph as a function of $\log(1/C_2)$. The experimental curve was shifted horizontally until it coincides with one of the theoretical curves. The extent of this shift gave $\log X$, and the coinciding theoretical curve gave the value for $\bar{u}/\bar{v}$. The values for X and $\bar{u}/\bar{v}$ derived in this way, for the membrane electrolyte systems are given in Table 3. The TMS method has been generally used and widely accepted for the evaluation of the charge density of a membrane.

The charge concept of TMS for charged membranes is a pertinent starting point for the investigation of the actual mechanisms of the ionic or molecular processes which occur in the membrane phase. The system considered for this study consists of an ionizable membrane of uniform thickness which separates two bulk solutions of a uni-univalent electrolyte of concentrations C_1 and C_2. It is assumed that the system is isothermal and no pressure is applied across the membrane. The ionizable groups are fixed on the polymer network present in the given membrane. The final expression [42] for the membrane potential is shown in equation (3) below:

$$\Delta\phi = -\left(\frac{RT}{F}\right)\left[\frac{1}{\beta}\ln\frac{C_2}{C_1}\left(1+\frac{1}{\beta}-2\alpha\right)\ln\frac{C_2+\alpha\beta\chi}{C_1+\alpha\beta\chi}\right] \qquad (3)$$

where R is the molar gas constant, T the absolute temperature and F the Faraday constant of the system, respectively. C_1 and C_2 are the concentrations of the solution.

Where

$$\alpha = \left(\frac{u}{u+v}\right) \qquad (4)$$

where α is related to membrane selectivity for ions. The symbols u and v represent the molar mobilities of positive and negative ions respectively and

$$\beta = 1 + \left(\frac{KF\chi}{u}\right) \qquad (5)$$

whereβ is related to the effect of adsorption of ions upon the membrane materials. K is the constant which is considered to depend upon the viscosity of the solution and the structural details of the polymer network of which the membrane is composed, χ is the charge density of the membrane. α and β parameters have been assumed to be independent of salt concentration.

Kobatake et al. [42] derived two useful limiting forms of equation (3). These are: a) when C_2 becomes sufficiently small with γ fixed; equation may be expanded to give

$$|\Delta\phi_r| = \frac{1}{\beta}\ln\gamma - \frac{\gamma-1}{\alpha\beta\gamma}\left(1+\frac{1}{\beta}-2\alpha\right)\frac{C_2}{\chi} \tag{6}$$

where γ is the concentration ratio $\left(\frac{C_1}{C_2}\right)$ and $|\Delta\phi_r|$ is the absolute value of a reduced membrane potential defined by

$$|\Delta\phi_r| = \frac{F\Delta\phi}{RT} \tag{7}$$

(b)Further Kobatake et al. [42]showed that at a fixed γ the inverse of an apparent transport number t_{-app} for the co-ion species in a negatively charged membrane is proportional to the inverse of the concentration C_2 in the region of high salt concentration. t_{-app} is defined by the relation

$$|\Delta\phi_r| = \left(1-2t_{-app}\right)\ln\gamma \tag{8}$$

The derived transport number value is called the apparent number i.e. t_{-app} because in this type of measurement water transport has not been taken into account. This apparent value will be close to the true value in dilute solutions. Substituting expressions (7) and (8) in equation (3) and expanding the resulting expression for $1/t_{-app}$ in powers of $1/C_2$ gives the following equation:

$$\frac{1}{t_{-app}} = \frac{1}{(1-\alpha)} + \frac{(1+\beta-2\alpha\beta)}{2(1-\alpha)^2 \ln \gamma}\alpha\left(\frac{\chi}{C_2}\right) + - - - - - \qquad (9)$$

Equation (6) indicates that, a value of β and a relation between α and χ can be obtained by evaluation of the intercept and the initial slope of a plot of $|\Delta\phi_r|$ against C_2 which is shown in Figure 8. The value of intercept is equal to $1/\beta \ln \gamma$ from which β is evaluated as shown in Table 4.

Equation (9) indicates that, the intercept of a plot of $1/t_{-app}$ against $1/C_2$ at fixed γ allows the value of α to be determined. Plots of $1/t_{-app}$ against $1/C_2$ for various uni-univalent electrolytes are shown in Figure 9. The value of intercept is equal to $1/(1-\alpha)$, from which α may be evaluated as shown in Table 4.

If this value of α is inserted in the relation obtained from the initial slope for $|\Delta\phi_r|$ against C_2, the desired value for χ_c can be determined. Once α and β are known in the manner described above, the values of χ_d may be evaluated from the initial slope for $1/t_{-app}$ against $1/C_2$.

Table 4. Derived values α and β for the nylon-6,6 nickel carbonate membrane

Electrolyte	α	β
KCl	0. 052	1.44
NaCl	0.048	1.32
LiCl	0.039	1.25

If the equation suggested by Kobatake for the membrane potential is correct, then the two values of χ i.e. χ_c and χ_d thus determined from the opposite limits should agree with one another. The values are given in Table 3 which are closed together thereby confirming the applicability of Kobatake's equation to these systems [43]. The charge density is higher in the region of low concentrations (χ_d) than in high concentrations (χ_c) due to the ionic radii.

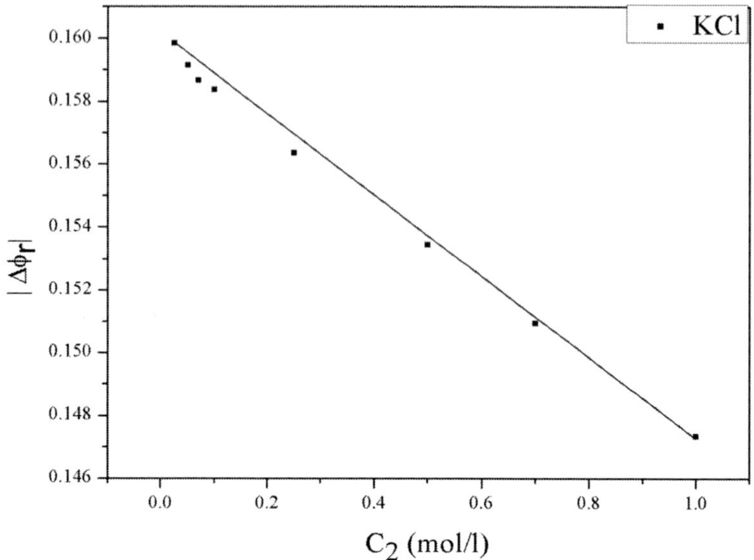

Figure 8. Plot of membrane potentials $\left|\Delta\phi_r\right|$ versus C_2 for nylon-6,6 nickel carbonate membrane using KCl electrolytes.

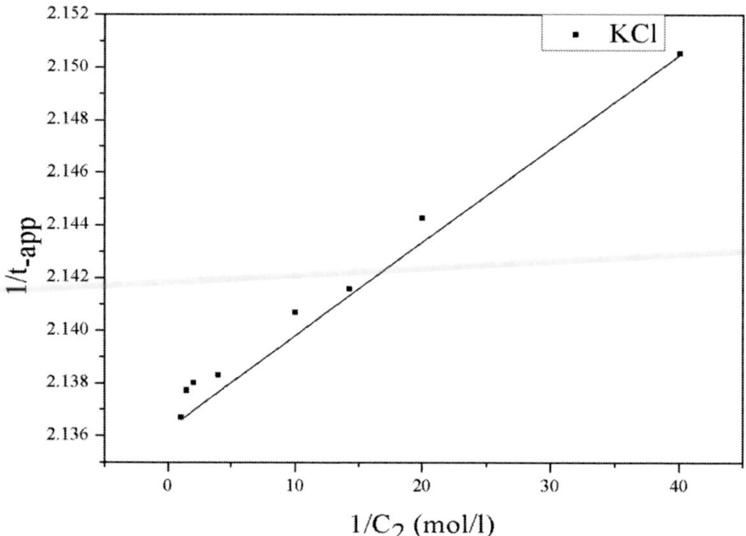

Figure 9. Plot of membrane potentials $1/t_{-app}$ versus $1/C_2$ for the nylon-6,6 nickel carbonate membrane using KCl electrolytes.

Recently, Nagasawa et al.   [44,45] based on the charge concept developed a theory for the membrane potential and used it for the evaluation of charge density of membranes. The Nagasawa's theory assumes the following: (a) the rate determining factor for the permeation of ions and water is the membrane phase, (b) the electrolyte solutions inside and outside of the membrane are in thermodynamic equilibrium on either side of the membrane, and (c) the effect of the stagnant layer on the permeation velocities is negligible for a thick membrane.

The total membrane potential $\Delta\phi$ was considered as the sum of diffusion potential ($\Delta\phi_d$) inside the membrane and the electrostatic potential difference ($\Delta\phi_e$) between the membrane surfaces and the electrolyte solutions on both sides of the membrane. The diffusion potential was obtained by integrating the basic flow equation for diffusion while the electrostatic potential was calculated from the Donnan's theory. In the absence of an externally applied electric field or a pressure gradient, at the limit of high electrolyte concentration (where the effect of flow of water is negligible), the expression for membrane potential can be written as:

$$ -\Delta\phi = \frac{RT}{F}\left(\frac{\gamma-1}{\gamma}\right)\left(\frac{\theta}{2}\right)\left(\frac{1}{C_2}\right) \tag{10} $$

where γ is the solution-concentration ratio (C_2/C_1) fixed at 10 , θ is the charge density on the membrane. Equation (10) predicts a linear relationship between $\Delta\phi$ and ($1/C_2$) in Figure 10. The values of θ derived from the slopes of the linear plots are given in Table 3.

All the theories derived for the charge membrane and used in these investigations gives the charge density. The charge densities for a 1:1 electrolyte solution were found to be KCl > NaCl > LiCl. Table 3 shows that the charge densities (X, χ_c, χ_d and θ) found by three theories are closer to each other. In other words, a slight difference in the values of the charge may be ascribed to different graphical procedures adopted for the evaluation.

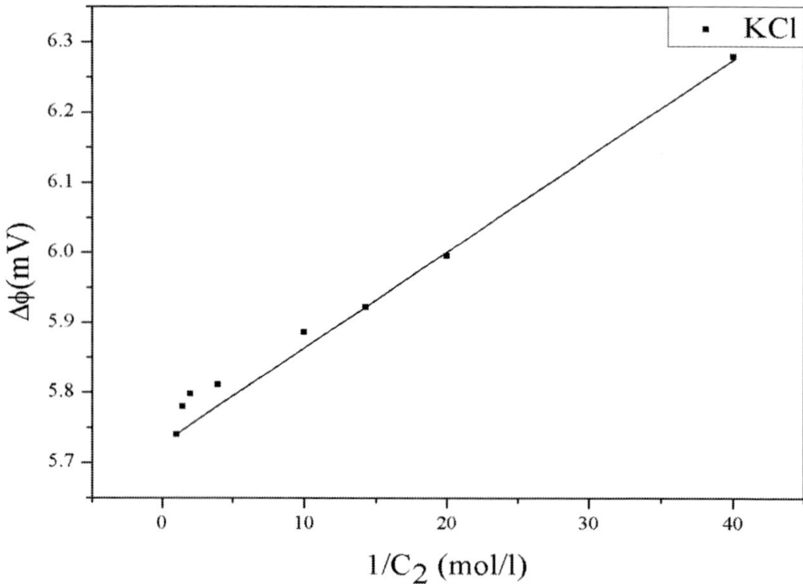

Figure 10. Plot of membrane potentials $\Delta\phi$ versus $1/C_2$ for the nylon-6,6 nickel carbonate membrane using KCl electrolytes.

In addition to the above equation (1), TMS further extended their theory and derived another equation for calculating the membrane potential. The extended equations of the TMS are the Donnan equation for the partition coefficients, ion concentrations at the surface between the membrane and the external solution, the Nernst-Planck equation, and the different electroneutrality conditions for the inside membrane and the external solutions [46].

$$\Delta\phi = \Delta\phi_{Don}(I) + \Delta\phi_{diff} + \Delta\phi_{Don}(II) \tag{11}$$

$$\Delta\phi = \Delta\phi_{Don} + \Delta\phi_{diff} \tag{12}$$

$$= -\frac{RT}{V_k F}\ln\left(\frac{\gamma''_{\pm} C_2 \overline{C}_{1+}}{\gamma'_{\pm} C_1 \overline{C}_{2+}}\right) - \frac{RT}{V_k F}\frac{\overline{\omega}-1}{\overline{\omega}+1}\times\ln\left(\frac{(\overline{\omega}+1)\overline{C}_{2+} + (V_x/V_k)X}{(\overline{\omega}+1)\overline{C}_{1+} + (V_x/V_k)X}\right) \tag{13}$$

Table 5. The calculated values of the parameters t_+, $\overline{U}$, $\overline{\omega}$, $K_\pm$, q, $\overline{C}_+$ and $\overline{C}_-$ of nylon-6,6 nickel carbonate membrane with different concentration of electrolytes using Equations 14 - 24

C_2 (mol/l)	t_+	$\overline{U}$	$\overline{\omega}$	$K_\pm$	q	$\overline{C}_+$	$\overline{C}_-$
KCl							
1	0.5486	0.0972	1.2155	0.9993	0.7812	0.7798	0.7817
0.7	0.5489	0.0979	1.2172	0.9991	0.7940	0.5544	0.5562
0.5	0.5498	0.0996	1.2213	0.9987	0.8311	0.4141	0.4160
0.25	0.5505	0.1010	1.2247	0.9975	0.8495	0.2109	0.2128
0.1	0.5511	0.1022	1.2276	0.9938	0.8802	0.0865	0.0884
0.07	0.5519	0.1038	1.2318	0.9911	0.8984	0.0614	0.0632
0.05	0.5525	0.1050	1.2348	0.9876	0.9167	0.0443	0.0462
0.025	0.5525	0.1064	1.2382	0.9752	0.9499	0.0222	0.0241
NaCl							
1	0.5493	0.0986	1.2188	0.9995	0.8063	0.8055	0.8066
0.7	0.5504	0.1008	1.2243	0.9994	0.8187	0.5722	0.5733
0.5	0.5511	0.1022	1.2276	0.9992	0.8369	0.4176	0.4187
0.25	0.5522	0.1044	1.2331	0.9985	0.8550	0.2129	0.2139
0.1	0.5528	0.1057	1.2365	0.9964	0.8847	0.0876	0.0887
0.07	0.5533	0.1067	1.2390	0.9948	0.9078	0.0626	0.0637
0.05	0.5541	0.1083	1.2429	0.9928	0.9252	0.0453	0.0464
0.025	0.5547	0.1094	1.2459	0.9856	0.9502	0.0228	0.0239
LiCl							
1	0.5504	0.1008	1.2243	0.9996	0.8247	0,8239	0.8249
0.7	0.5512	0.1025	1.2285	0.9995	0.8368	0.5850	0.5860
0.5	0.5520	0.1040	1.2323	0.9993	0.8546	0.4265	0.4275
0.25	0.5528	0.1057	1.2365	0.9986	0.8666	0.2158	0.2168
0.1	0.5536	0.1072	1.2403	0.9966	0.8903	0.0882	0.0892
0.07	0.5546	0.1093	1.2454	0.9951	0.9132	0.0631	0.0641
0.05	0.5557	0.1115	1.2510	0.9932	0.9305	0.0457	0.0467
0.025	0.5565	0.1130	1.2549	0.9864	0.9552	0.0230	0.0240

In the above equation R, T and F have their usual significance, $\gamma'_\pm$ and $\gamma''_\pm$ are the mean ionic activity coefficients, $\overline{\omega} = \dfrac{\overline{u}}{\overline{v}}$ is the mobility ratio of the cation to the anion in the membrane phase, and $\overline{C}_{1+}$ and $\overline{C}_{2+}$ are the cation concentrations on opposite sides of the membrane phase. Here V_k and V_x refer to the valency of the cation and fixed-charge group on the membrane matrix, respectively. The internal (membrane) counter-ion concentration is given by equation (14):

$$\overline{C}_+ = \sqrt{\left(\frac{V_x X}{2V_k}\right)^2 + \left(\frac{\gamma_\pm C}{q}\right)^2} - \frac{V_x X}{2V_k} \tag{14}$$

and co-ion concentration is given by equation (15):

$$\overline{C}_- = \sqrt{\left(\frac{V_x X}{2V_k}\right)^2 + \left(\frac{\gamma_\pm C}{q}\right)^2} + \frac{V_x X}{2V_k} \tag{15}$$

Here q is the charge effectiveness of the membrane and is defined by equation (16):

$$q = \sqrt{\frac{\gamma_\pm}{K_\pm}} \tag{16}$$

where $K_\pm$ is the charge distribution coefficient.

The condition of electro-neutrality in the membrane and the external solution is expressed in the following equations:

$$\sum_i (Z_i C_i + Z_x X_x) = 0 \tag{17}$$

$$\sum_i |Z_i C_i| = \sum_x |Z_x X_x| \tag{18}$$

where C_i is the ith ion concentration with the charge Z_i of the external solution and X_x is the xth ion concentration with the charge Z_x in the membrane phase. $K_{\pm}$, the charge distribution coefficient, is expressed as follows:

$$K_{\pm} = \sum \frac{\overline{C}_i}{|Z_i C_i|} \tag{19}$$

$$\sum \overline{C}_i = \sum |Z_i C_i - Z_x X_x| \tag{20}$$

$$X = \sum Z_x X_x \tag{21}$$

where $\overline{C}_i$ is the ith ion concentration in the membrane phase.

The Donnan exclusion of co-ions at low concentration, which is associated to neutral or weak charge of membrane [47,48], but at a high concentration, the Donnan effect can be neglected and the membrane potentials are mainly due to diffusion of counter and co-ions through the membrane; thus, $\Delta\phi \approx \Delta\phi_{diff}$. For 1:1 electrolytes, it is expressed by

$$\Delta\phi \approx \Delta\phi_{diff} = \left(\frac{RT}{F}\right)\left[\frac{t_+}{|Z_+|} - \frac{t_-}{|Z_-|}\right]\ln\left(\frac{C_2}{C_1}\right) \tag{22}$$

$$\Delta\phi = \left(\frac{RT}{F}\right)U \ln\left(\frac{C_2}{C_1}\right) \tag{23}$$

$$\frac{t_+}{t_-} = \frac{\overline{u}}{\overline{v}} \tag{24}$$

where t_+ and t_- are the cation and anion transport numbers in the membrane, Z_+ and Z_- their valencies, respectively of the system and U is a parameter related to the combined transport numbers of the ions in the membrane. The transport number of the ion i in the membrane represents the amount of current transported for one ion with respect to the total current crossing the membrane, $t_i = \dfrac{I_i}{I_T}$, and for single salts $t_+ + t_- = 1$.

The values of $\gamma_\pm$ were the usual charted values for electrolytes. The values of the parameters t_+, $\overline{\omega}$, U, $K_\pm$, q, $\overline{C}_+$ and $\overline{C}_-$ are listed in Table 5.

From Table 5, it shows that the calculated transport numbers are low at higher concentration. Transport number decreases with an increase of external electrolyte concentration [19]. The transport values for the electrolytes follow the sequence for the cation in the order $Li^+ > K^+ > Na^+$, which described elsewhere [49].

CONCLUSION

The preparation of nylon-6,6 nickel carbonate membrane at different compositions of material were described. In order to understand the mechanism of transport of ions, the membrane potential measurements were carried out using different concentrations of 1:1 electrolyte (KCl, NaCl and LiCl) solutions and also to evaluate various membrane parameters controlling the transport phenomena. The observed potential data increased with an increase in concentration of all the verified 1:1 electrolyte solutions and followed the sequence KCl < NaCl < LiCl. The TMS, Kobatake and Nagasawa's methods were used for the estimation of thermodynamical charge density of membranes. The charge densities for 1:1 electrolye were found to be KCl > NaCl > LiCl. The value of charge density thus obtained are depending on the concentration of external solution and on the type of electrolyte species. In the present investigation, the mean value of charge density obtained by using different equations of Kobatake is found to be closer to that obtained by Nagasawa and TMS theory. The charge densities evaluated from various procedures are not much different from each other. A slight difference in the values of charge density may be ascribed to different graphical procedures adopted for the evaluation. The membrane potential was quite sound and that their use for the evaluation of charge density of membrane was justified for at least the systems under investigation. The order of transport numbers of cation for 1:1 electrolyte solution was found to be KCl < NaCl < LiCl.

REFERENCES

[1]   Khan, A.A.; Paquiza, L. *Desalination.* 2011, 272, 278-285.

[2]   Zimmerman, J.; Mark, H.F.; Bikales, N.M. *Encyclopedia of polymerscience and engineering*; Wiley: New York, 1988; Vol.6, pp 802-839.

[3]   King, J.A.; Tucker, K.W.; Vogt, B.D.; Weber, E.H.; Quan, C. *Polym.Compos.* 1999, 20, 643-654.

[4]   Levon, K.; Nasybulin, E.; Menshikova, I.; Sergeyev, V.; Zezin, A. *Polym. Sci.* A. 2009, 51, 701-707.

[5]   Kang, E.; Kim, M.; Oh, J.S.; Park, D.W.; Shim, S.E. *Macromol. Res.* 2012, 20, 372-378.

[6]   Rinken, T.; Järv, J.; Rinken, A. *Anal. Chem.* 2007, 79, 6042-6044.

[7]   Nirmala, F.R.; Jeong, J.W.; Oh, H.J.; Navamathavan, R.; El-Newehy, M.; Al-Deyab, S.S.; Kim, H.Y. *Fiber. Polym.* 2011, 12, 1021-1024.

[8]   El-Newehy, M.; Al-Deyab, S.; Kenawy, E.R.; Abdel-Megeed, A. J. *Nanomaterials.* 2011, 2011, 1-8.

[9]   Wallner, H.; Preis, W. Gamsjäger, *Thermochim. Acta.* 2002, 382, 289-296.

[10]   Kohls, D.W.; Rodda, J.L. *Am. Mineral.* 1996, 51, 677-684.

[11]   Berezina, N.P.; Kinonenko, N.A.; Dyomina, O.A.; Gnusin, N.P. Adv. *Colloid Interface Sci.* 2008, 139, 3-28.

[12]   Le, X.T. *J.Colloid Interface Sci.* 2008, 325, 215-222.

[13]   Takagi, R,; Nakagaki, M. J. *Membr. Sci.* 1990, 53, 19.

[14]   Hsu, J.-P.; Kuo, Y.-C. J. *Colloid Interface Sci.* 1995, 176, 256.

[15]   Takagi, R.; Nakagaki, M. J. *Membr. Sci.* 1996, 111, 19.

[16]   Abrabri, M.; Rafiq, M.; Larbot, A.; Persin, M.; Sarrazin, J.; Cot, L. *J.Membr. Sci.* 1998, 139, 275.

[17]   Khalid, M.; Mohammad, F. *Synthesis Metals.* 2009, 159, 119-122.

[18]   Arfin, T.; Rafiuddin. *Desalination.* 2012, 284, 100-105.

[19]   Arfin, T.; Rafiuddin, *Electrochim. Acta.* 2011, 56, 7476-7483.

[20]   Arfin, T.; Rafiuddin, *Electrochim. Acta.* 2010, 55, 8628-8631.

[21]   Arfin, T.; Yadav, N.; *Anal. Bioanal. Electrochem.* 2012, 4, 135-152.

[22]   White, W.B. In *The carbonate minerals*; Farmer, V.C.; Ed.; The infrared spectra of minerals; Mineralogical Society: London, 1974.

[23]   Reddy, B.J.; Frost, R.L. NeuesJahrbuchfuerMineralogie. *Monatshefte* 2004, 11, 525-536.

[24]   Li, J.; Yan, R.; Xiao, B.; Liang, D.T.; Lee, D.H. *Energy Fuels* 2008a, 22, 16-23.

[25] Nair, S.S.; Ramesh, C. *Macromolecules.* 2005, 38, 454-462.
[26] Jakes, J.;Krimm,S. Spectrochim. Acta. 1971, 27A, 35-63.
[27] Ishisue, T.; Okamoto, M.; Tashiro, K. Polymer. 2010, 51, 5585-5591.
[28] Khan, A.A.; Akhtar, T. Electrochim. Acta. 2009, 54, 3320-3329.
[29] Haber, J. Pure and Applied Chemistry. 1991, 63, 1227-1246.
[30] Schmid, G.; Schwarz, H. Z. Electrochem. 1951a, 55, 295-307.
[31] Arfin, T.; Rafiuddin. *Electrochim. Acta.* 2009, 54, 6928-6934.
[32] Arfin, T.; Jabeen, F.; Kriek, R.J. *Desalination.* 2011, 274, 206-211.
[33] Westermann-Clark, G.B.; Christoforou, C.C. *J. Electroanal. Chem.* 1986, 198, 213-231.
[34] Teorell, T. *Proc. Soc. Exp. Bio.* 1935, 33, 282-285.
[35] Teorell, T. *Proc. Natl. Acad. Sci. USA.* 1935, 21, 152-161.
[36] Meyer, K.H.; Sievers, J.-F. *Helv. Chim. Acta.* 1936, 19, 649-664.
[37] Meyer, K.H.; Sievers, J.-F. *Helv. Chim. Acta.* 1936, 19, 665-677.
[38] Meyer, K.H.; Sievers, J.-F. *Helv. Chim. Acta.* 1936, 19, 987-995.
[39] Lakshminarayanaiah, N. *Transport Phenomena in Membranes*; Academic Press: New York, 1969, 202-203.
[40] Beg, M.N.; Altaf, I. *J. Appl. Polymer Sci.* 1990, 39, 1495-1506.
[41] Beg, M.N.; Matin, M.A. *J. Membr. Sci.* 2002, 196, 95-102.
[42] Kobatake, Y.; Takeguchi, N.; Toyoshima, Y.; Fuita, H. J. *Phys. Chem.* 1965, 69, 3981-3988.
[43] Kamo, N.; Oikawa, M.; Kobatake, Y. *J. Phys. Chem.* 1973, 77, 92-95.
[44] Nagasawa, M.; Kagawa, I. *Discuss. Faraday Soc.* 1956, 21, 52-60.
[45] Tasaka, M.; Aoki, N.; Konda, Y.; Nagasawa, M. *J. Phys. Chem.* 1975, 79, 1307-1314.
[46] Arfin, T.; Rafiuddin, *J. Electroanal. Chem.* 2009, 636, 113-122.
[47] Vázquez, M.I.; Lara, R.; Gálan P.; Benavente, J. *J. Membr. Sci.* 2005, 256, 202-208.
[48] Cañas, A.; Ariza, M.J.; Benavente, J. *J. Colloid Interface Sci.* 2002, 246, 150-156.
[49] Jabeen, F. Rafiuddin, J. Sol-Gel Sci. *Technol.* 2007, 44, 195-202.

In: Halides
Editor: Jean Lefebure

ISBN: 978-1-62417-947-1
© 2013 Nova Science Publishers, Inc.

Chapter 3

NUCLEOPHILIC SUBSTITUTION AND PALLADIUM-CATALYZED COUPLING MECHANISMS OF AROMATIC HALIDES

Linjun Shao, Kai Chen, Minfeng Zeng, Chen Wang, Chenze Qi and Xian-Man Zhang[*]

Department of Chemistry, University of Shaoxing, Shaoxing, Zhejiang Province, People's Republic of China

ABSTRACT

Unactivated aromatic halides typically do not react with nucleophiles unless submitted to extremely drastic reaction conditions. The extremely low reactivities of aromatic halides can be attributed to their sTable aromatic π electron structure. But the unactivated aromatic halides can readily undergo radical nucleophilic substitution ($S_{RN}1$) and palladium-catalyzed reactions under relatively mild conditions. The remarkable enhancement of the aromatic halide reactivities has been attributed to their activation by electron transfer and/or palladium oxidative insertion. The palladium-catalyzed coupling of aromatic halides has been become as one of the most important chemical transformations for the construction of carbon-carbon chemical bonds in synthetic organic chemistry. The related reaction mechanisms and their synthetic

---

[*] E-mail: xian-man.zhang@usx.edu.cn and xianmanzhang@yahoo.com.

applications will be discussed with an emphasis on the author's research interests.

INTRODUCTION

Aromatic halides are generally referred to as organic chemical compounds in which at least one or more aromatic carbon atoms are linked by covalent bonds with one or more halogen atoms. Undoubtedly, they are among the most studied chemical compounds in all fields of chemistry, resulting in many widely-accepted chemistry concepts.

Aromatic halides are not only naturally produced in massive amounts by microorganisms, [1] but are also the important starting materials, intermediates and/or final products for many chemical industries such as pharmaceuticals, dyes, polymers, pesticides, electronic chemicals, etc. [?] Although their preparation, properties, reactions and applications are well reviewed and documented in the literature, [3] many negative effects of organohalides have been discovered including carcinogen, harmfully environmental impacts and ozone depletion. [2, 3] In the past several decades, tremendous progress has been made for the palladium-catalyzed reactions of aromatic halides. [4] Thus, it is not surprising why research on aromatic halides is not diminishing, but instead continues to flourish.

***Nucleophilic Substitution of Aliphatic and Aromatic Halides*.** Beginning with the groundbreaking studies by C.K. Ingold in 1930s, nucleophilic substitution concept has been developed from the chemical reactions of aliphatic halides and it has become one of the most important concepts in modern organic chemistry. [5] The nucleophilic substitution of primary alkyl halides generally proceed through a one-step bimolecular mechanism (S_N2) as shown in Scheme 1 and the rate-determining step is the formation of the transition state. The departure of the halide leaving group occurs simultaneously with the nucleophile backside attack, resulting in a predicTable stereochemical configuration of the final substitution products.

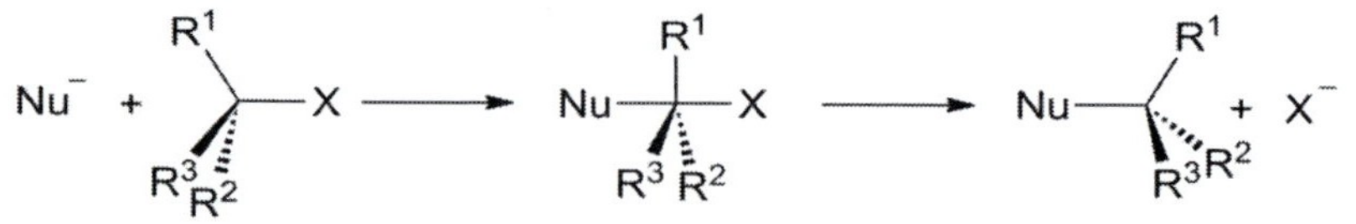

Scheme 1.

The steric hindrance of the secondary and tertiary alkyl halides will impede or even completely prevent the nucleophile backside attack. As a result, it is difficult for the secondary and tertiary alkyl halides to proceed through the S_N2 mechanism, but instead undergo a two-step unimolecular nucleophilic substitution mechanism (S_N1) as shown in Scheme 2. The rate-determining step is the heterolytic cleavage of the C-X bond to form a planar carbocation intermediate, yielding the racemized products because the nucleophile can attack the planar intermediate on either side.

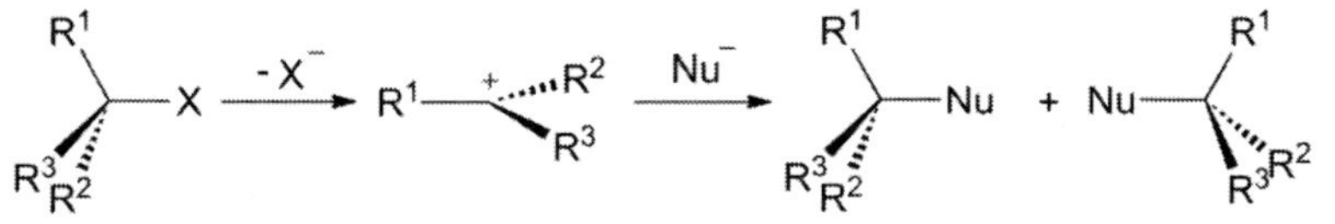

Scheme 2.

On the contrary, the S_N1 and S_N2 mechanisms are not common for the nucleophilic substitution of aromatic halides. The S_N2 reaction requires the backside attack, implying that the nucleophile must appear inside the aromatic ring in order to attack from the backside of the C_{sp^2}-X bond. The aromatic halides also hardly undergo the S_N1-type nucleophilic substitution primarily due to the relatively stronger C_{sp^2}-X bond strengths, which is supported by theoretical calculations of the carbon-halogen bond dissociation enthalpies (Table 1) for both aliphatic and aromatic halides. Examination of Table 1 shows that both homolytic and heterolytic dissociation enthalpies of the aromatic C_{sp^2}-X bonds are considerably stronger than the corresponding aliphatic C-X bonds, especially in the polar media such as water and dimethyl sulfoxide (DMSO). The much stronger C_{sp^2}-X bond strength is clearly associated with a combination of the aromatic carbon sp^2 hybridization and the partial double bond feature (hyperconjugation) between the halogen atom with the aromatic π system. Moreover, the C_{sp^2}-X bond cleavage will form an unsTable phenyl carbocation with the positive charge primarily localized on the sp^2 carbon orbital. The phenyl carbocation has no delocalization into the aromatic π system because its orbital is orthogonal to the aromatic π structure.

Benzyne Mechanism. Unactivated aromatic halides can be forced to undergo nucleophilic substitution when submitted to drastic reaction conditions such as extremely high temperature and pressure. For example, phenol can be obtained from the reaction of chlorobenzene with concentrated sodium hydroxide *only* at temperatures over 350 ^{0}C and high pressure, or with steam at 400~450 ^{0}C over calcium phosphate catalyst as described in the Dow Chemical patent. [6] Similarly, aniline can *only* be obtained from the reaction of chlorobenzene with ammonia under high temperature and pressure as well as copper catalyst. [7]

Table 1. Homolytic and Heterolytic Bond Dissociation Enthalpies of the Carbon- Halogen Bonds in Aliphatic and Aromatic Halides[a]

Halides	$BDE_{C-X}{}^{b}$	$BDE_{C-X}{}^{c,d}$	$BDE_{C-X}{}^{c,e}$	$BDE_{C-X}{}^{c,f}$
Me$_2$CH-Cl	83.5	162.1	33.7	34.2
Me$_3$C-Cl	85.2	145.9	22.2	22.4
Ph-Cl	97.2	194.7	73.2	73.5
Me$_2$CH-Br	68.8	151.9	35.9	36.3
Me$_3$C-Br	70.2	135.3	23.9	24.1
Ph-Br	81.7	183.7	74.9	75.2
Me$_2$CH-I	61.2	153.2	26.2	27.6
Me$_3$C-I	61.9	136.1	13.8	15.0
Ph-I	74.3	185.3	64.8	65.9

[a] In kcal/mol, calculated at the M06-2X/6-311++G(2df,2p)//B3LYP/6-31G(d) level using SMD as the solvent model; [b] Homolytic cleavage in the gas phase; [c] Heterolytic cleavage to form halide anion (Scheme 2); [d] In the gas phase; [e] In the water solution; [f] In the DMSO solution.

However, instead of proceeding through the addition/elimination mechanism, the unactivated aromatic halides undergo via an unsTable reactive intermediate (benzyne). When isotope-labeled chlorobenzene was employed to react with potassium amide, the isotope-labeled carbons were equally distributed at positions 1 and 2 of the final aniline products, leading to the proposal of the benzyne mechanism (Scheme 3). [8] The benzyne intermediate contains an unsTable triple bond from the overlap of the adjacent sp^2 orbitals, which have no overlap with the aromatic π orbitals. The entering nucleophile can attack either aromatic carbon of the triple bond, resulting in nonregiospecific substitution products. Nevertheless, the reactive intermediate (benzyne) has been trapped through many Diels-Alder reactions. [9]

Scheme 3.

S_NAr Mechanism. Aromatic halides can readily undergo the direct nucleophilic substitution (S_NAr) if the aromatic ring contains one or more strong electron-withdrawing nitro groups at the *ortho* and/or *para*-positions. The S_NAr mechanism is characterized by the formation of a Meisenheimer complex, followed by the loss of the halide anion. The nitro group(s) can stabilize the Meisenheimer complexes because of the efficient delocalization of the negative charge as shown in Scheme 4. [10]

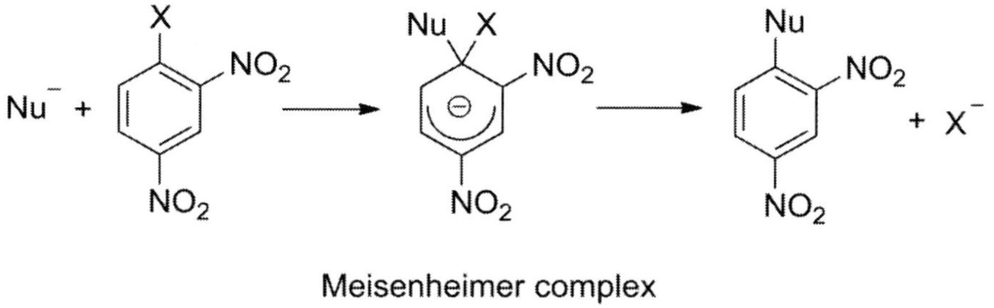

Meisenheimer complex

Scheme 4.

Radical Nucleophilic Substitution of Aromatic Halides ($S_{RN}1$). The unactivated aromatic halides can readily undergo the radical nucleophilic substitution under relatively milder reaction conditions. In 1970, Bunnett et al. observed that the reactions of 5-iodo- and 6-iodo-1,2,4-trimethylbenzenes with potassium amide gave different substitution products. [11] These two reactions should afford the same products because of the identical reactive intermediate if they proceed via the benzyne mechanism (Scheme 3). In contrast, the major amination products were found to be those derived from the corresponding *ipso* substitutions, i.e. the products by amide anion attack on the same carbon from which iodine departs. But addition of radical scavengers such as tetraphenylhydrazine shifts the amination products similar to those expected from the benzyne mechanism. Therefore, a chain radical unimolecular nucleophilic substitution mechanism ($S_{RN}1$) has been proposed as shown in Scheme 5. [12] Similar chain radical mechanism has also been independently proposed for the aliphatic halides such as *p*-nitrocumyl halides [13] and halonitropropanes. [14] A wide variety of aromatic halides have been shown as suiTable substrates for the $S_{RN}1$ reactions and some unique synthetic approaches have been developed based on the $S_{RN}1$ reactions. [12, 15]

Initiation

$$ArX + e \longrightarrow ArX^{\cdot-}$$

$$ArX^{\cdot-} \longrightarrow Ar\cdot + X^-$$

Propagation

$$Ar\cdot + Nu^- \longrightarrow ArNu^{\cdot-}$$

$$ArX + ArNu^{\cdot-} \longrightarrow ArNu + ArX^{\cdot-}$$

$$ArX + Nu^- \longrightarrow ArNu + X^-$$

Scheme 5.

$S_{RN}1$ reactions are generally believed to proceed via the chain mechanism, but a non-chain radical mechanism has also been proposed. [16] We have demonstrated that the nucleophilic substitutions of o- and p-nitrohalobenzenes with ethyl α-cyanoacetate carbanion proceed via a non-chain radical mechanism in DMSO solution. [17]

Activation of Aromatic Halides by Electron transfer. We may wonder why the nucleophilic substitution of the unactivated aromatic halides can proceed under much milder reaction conditions through the radical and no the polar mechanism. The remarkable difference in reactivity is associated with the activation of the sTable aromatic halide substrates by electron transfer because formation of radical anion can dramatically reduce the bond strength. [18]

Homolytic bond dissociation enthalpies (BDEs) of various acidic H-A bonds have been determined from the equilibrium acidities and the oxidation potentials of their conjugate anions. [19] The acidic H-A BDEs in many radical anions have also been determined by an equation derived from the thermodynamic cycle as shown in Scheme 6. [18a] Table 2 presents several acidic C-H and N-H BDEs in both neutrals and radical anions. Examination of Table 2 shows that addition of one electron weakens the corresponding acidic H-A bond strength by about 30~60 kcal/mol!

$$H\text{-}A^{\cdot-} \;\rightleftharpoons\; H\text{-}A \;+\; e^- \qquad E_{re}(HA)$$
$$H\text{-}A \;\rightleftharpoons\; H^+ \;+\; A^- \qquad pK_{HA}$$
$$H^+ \;+\; e^- \;\rightleftharpoons\; H^{\cdot} \qquad C$$
$$H\text{-}A^{\cdot-} \;\rightleftharpoons\; H^{\cdot} \;+\; A^- \qquad BDE_{HA^{\cdot-}}$$

Scheme 6.

Table 2. Weakening Effects on the Acidic H-A Bond Strength by Formation of Radical Anions [18, 20]

Substrates	BDE_{HA} [a] (in kcal/mol)	$BDE_{HA^{\cdot-}}$ [b] (in kcal/mol)	ΔBDE [c] (in kcal/mol)
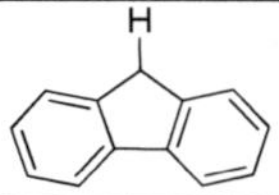 9H-fluorene	79.5	35	44.5
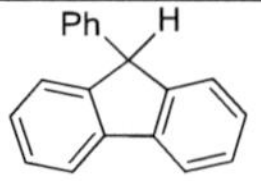 9-phenyl-9H-fluorene	74	28	46
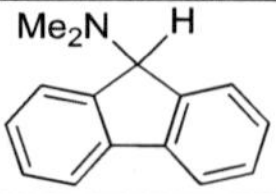 9-(dimethylamino)-9H-fluorene	71.5	38	33.5
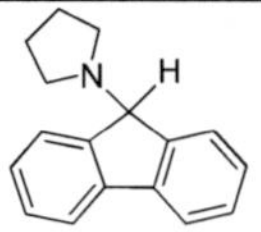 9-(pyrrolidin-1-yl)-9H-fluorene	72	38	34
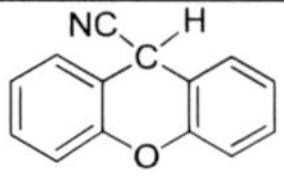 9H-xanthene-9-carbonitrile	69	26	43
 4-aminobenzonitrile	95	36	59
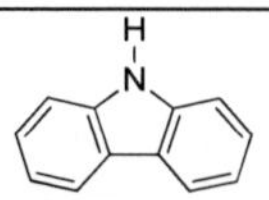 9H-carbazole	94	30	64

[a] Homolytic bond dissociation enthalpies in neutral molecules. [b] Homolytic bond dissociation enthalpies in radical anions. [c] Differences of the homolytic bond dissociation enthalpies between the neutrals and their radical anions.

Reduction of aliphatic and aromatic halides leads even much more severe bond weakening effects on the C-X bonds, primarily due to the relatively higher electronegativity of halogen than carbon atoms. [21, 22] As a result, the C-X anti-bonding orbital is much closer to the carbon rather than the halogen atom orbital as shown in Scheme 7. Interestingly, some halide radical anions are even thermodynamically unsTable because their C-X bond dissociation enthalpies are negative, suggesting that the C-X bond will immediately dissociate as soon as the capture of the extra electron. [20, 21] These explain why many extremely slow aromatic halide reactions by the polar mechanisms can readily undergo the $S_{RN}1$ mechanisms.

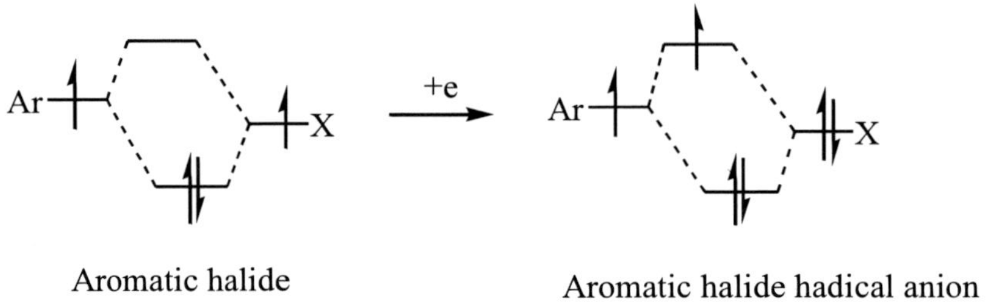

Aromatic halide   Aromatic halide hadical anion

Scheme 7.

Transition Metal-Catalyzed Coupling of Aromatic Halides. Traditionally, aromatic halides have limited applications for the construction of carbon-carbon and carbon-heteroatom bonds in synthetic organic chemistry because the aromatic nucleophilic substitution either relies on extremely high reaction temperature and pressure, or occurs under the highly reducing conditions of the $S_{RN}1$ reactions, or provide nonregiospecific substitution products through the benzyne mechanism. Moreover, all of these reactions lack the general tolerance for most common functional groups. [23] In other words, these aromatic reactions are not optimum for the organic synthetic applications.

In the past several decades, transition metals (especially palladium) catalyzed reactions have revolutionized the fundamental carbon-carbon and carbon-heteroatom bond transformations in synthetic organic chemistry. These reactions have the advantage of being simple to carry out under much milder conditions and not requiring anhydrous conditions or any special techniques. They are even stereospecific and regiospecific in addition to the tolerance of nearly all chemical functional groups. [4, 24] Indeed, the palladium-catalyzed reactions have become routine tools for the preparation of fine chemicals and pharmaceutically active compounds both in academic studies and industrial productions.

Oxidative Insertion of Zero-valent Palladium Species into Aromatic Halide. Transition metals usually have more than one oxidation state due to their partially filled *d*-orbitals, and their catalytic activities are generally dependent upon the kinetic barriers for switching between their oxidation states. [25] The oxidation state exchange will simultaneously induce electron transfer of the substrates, thereby dramatically enhancing their reactivities.

Heck, Suzuki, Sonogashira, Negishi, Stille and Ullmann couplings are few among the prominent palladium catalyzed reactions involving aromatic halides with different coupling reagents. [4] The Heck reaction is the palladium catalyzed cross-coupling of aromatic halides with alkenes; the Sonogashira reaction is that with alkynes; the Suzuki reaction is that with boronic acids; the Negishi reaction is that with organozinc reagents, and the Stille reaction is that with organotin reagents. But the palladium-catalyzed Ullmann reaction is referred to as the homocoupling of aromatic halides. [26, 27] Although these palladium mediated reactions have different reactants and products, the rate-determining step is all the same, i.e. the oxidative insertion of the zero-valent palladium species into the $C(sp^2)$-halogen bond to form the σ-arylpalladium halide reactive intermediate (Scheme 8). [28] The zero-valent palladium species is unsTable because it can easily lose its catalytic activity by aggregation and/or oxidation. Thus, the zero-valent palladium species (Pd^0) is generally formed *in situ* by reduction from the relatively sTable divalent Pd^{2+} compounds with suiTable hydrogen-donor and/or electron sources, [29] including formate salt, [30] hydrogen gas, [31] amines, [32] alcohols, [33] zinc, [34] triphenylarsine, [35] etc.

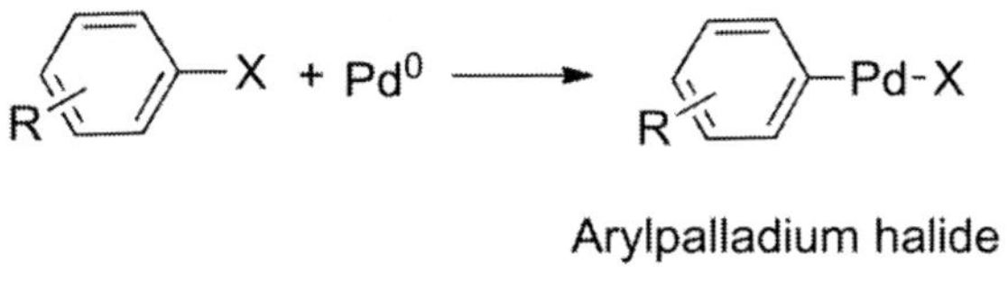

Scheme 8.

The oxidative insertion reactivity is highly dependent upon the carbon-halogen bond strength. [28] For example, the palladium catalyzed coupling reactions typically work well for aromatic iodides and bromides, but are generally sluggish for aromatic chlorides and fluorides because the C-Cl and C-F bonds are much stronger than those of the C-I and C-Br bonds. [36] The σ-arylpalladium halide intermediate can be regenerated from the catalytic cycle reactions for the Heck, Suzuki, Negishi, Kumada and Sonogashira reactions, suggesting that only a catalytic amount of the external reducing reagents is needed to regenerate the zero-valent palladium (0) species to complete the catalytic cycle for these reactions. *Nevertheless, a stoichiometric amount of the external reducing reagents is required for the Ullmann's homocoupling reactions.*

Palladium-Catalyzed Ullmann Reductive Homocouplings in DMSO and Alcoholic Media. The copper-mediated Ullmann reductive homocoupling reaction of aromatic halides has traditionally been used to synthesize symmetrical biaryls for over a century, [37, 38] but this reaction typically requires both high temperature (over 200 °C) and the consumption of a stoichiometric amount of copper. [39, 40] In comparison, the palladium catalyzed reductive homocoupling of aromatic halides can be carried out under much milder reaction conditions together with broader substituent tolerance. Indeed, it has been considered to be an attractive alternative to the traditional copper-mediated Ullmann reaction. [26, 27] The palladium catalyzed protocol still, however, requires a stoichiometric amount of the external reducing reagent to regenerate the zero-valent palladium species. Otherwise, a stoichiometric amount of the expensive palladium metal is required. [41]

Certainly, it is inconvenient and uneconomic to use a stoichiometric amount of the external reducing reagent. Furthermore, the removal of the oxidized products of the external reducing reagents may not be straightforward. We have demonstrated that various biphenyls can readily be synthesized in excellent yields from the palladium-catalyzed homocoupling of the corresponding aromatic halides in DMSO solution without any external reducing reagents (Scheme 9) and some of the representative results are summarized in Table 3. [42]

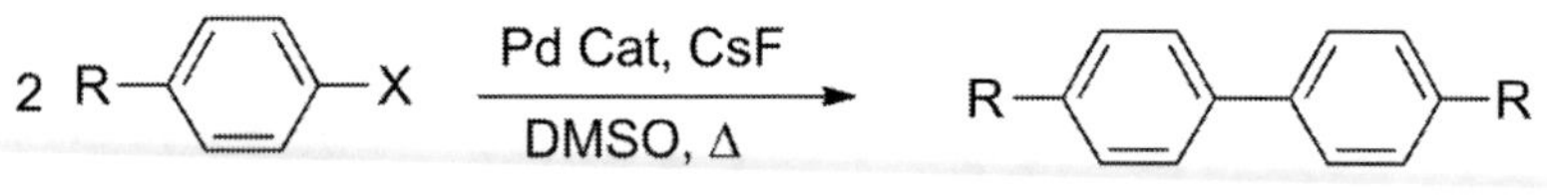

$$2 \; R{-}\langle\text{aryl}\rangle{-}X \xrightarrow[\text{DMSO, } \Delta]{\text{Pd Cat, CsF}} R{-}\langle\text{aryl}\rangle{-}\langle\text{aryl}\rangle{-}R$$

X = I and Br

Scheme 9.

A series of palladium catalysts were tested for the reductive homocoupling of aromatic halides in DMSO solution and Pd(dppf)Cl$_2$ was found to be the most effective among the screened palladium catalysts. The superior Pd(dppf)Cl$_2$ catalytic activities can be attributed to its unique electrochemical properties of the ferrocenyl structure, which can facilitate the switching of oxidation states for the palladium species. [43] The X-ray photoelectron spectroscopic (XPS) studies have confirmed the formation of the zero-valent palladium species (Pd^0) from the reduction of the oxidative Pd^{2+} species [Pd(dppf)Cl$_2$] in the absence of any external reducing reagents, suggesting that the solvent DMSO molecules are involved with the regeneration of the zero-valent palladium species from the reduction of the divalent palladium (Pd^{2+}) species. [42]

Table 3. Pd(dppf)Cl$_2$-Catalyzed Reductive Homocoupling of Various Aromatic Halides in DMSO Solution[a]

Aromatic Halides	Temperature (^{o}C)	Time (hour)	Biaryl yield (%)[b]
C_6H_5I	120	5	100 (95)
$4\text{-}CH_3C_6H_4I$	120	8	98.8 (94)
$4\text{-}CH_3OC_6H_4I$	120	8	92.7 (89)
$4\text{-}FC_6H_4I$	100	8	100 (96)
$4\text{-}ClC_6H_4I$	100	5	98.4 (95)
$4\text{-}O_2NC_6H_4I$	100	12	100 (96)
$4\text{-}EtO_2CC_6H_4I$	100	5	92.6 (91)
C_6H_5Br	120	10	100 (96)
$2\text{-}CH_3C_6H_4Br$	140	44	94.0 (90)
$4\text{-}CH_3C_6H_4Br$	140	44	84.2 (81)
$4\text{-}CH_3OC_6H_4Br$	140	44	74.2 (71)
$2\text{-}FC_6H_4Br$	100	1.5	94.3 (90)
$4\text{-}FC_6H_4Br$	120	10	100 (98)
$4\text{-}ClC_6H_4Br$	140	1.5	95.8 (93)
$4\text{-}CF_3C_6H_4Br$	100	5	100 (97)
$2\text{-}BrC_5H_4N$[c]	120	8	100 (96)
$2\text{-}Me\text{-}5\text{-}BrC_5H_3N$[d]	120	8	95.0 (91)
$2\text{-}MeO\text{-}5\text{-}BrC_5H_3N$[e]	120	8	92.3 (89)

[a] Reaction conditions: 1.0 mmol aromatic halide and 0.10 mmol Pd(dppf)Cl$_2$ catalyst; [b] Biaryl yields were determined from the GC/MS analysis. The isolated yields are listed in the parentheses; [c] 2-Bromopyridine; [d] 2-Methyl-5-bromopyridine; [e] 2-Methoxy-5-bromopyridine.

The palladium-catalyzed reductive homocoupling of aromatic halides can also be carried out in DMF [44] or alcoholic media without the need for any external reducing reagents. [45] The X-ray photoelectron spectroscopic studies (Figure 1) provided the conclusive evidence to show the formation of the reductive palladium (Pd^0) species from the reduction of the divalent palladium species (Pd^{2+}) in 3-pentanol solution. [45] Ethanol may also be involved with the regeneration of the zero-valent palladium species because it can significantly accelerate the Ullmann reductive homocoupling of aromatic halides when catalyzed by the carbon-supported palladium catalyst. [46] The electron binding energy peaks (338.0 and 343.2 eV) (Figure 1A) can be assigned to the palladium 3d electron of the divalent Pd^{2+} species. [47] The electron energy peaks of the palladium isolated from the reaction mixture is shown in Figure 1C, in which the palladium 3d electron binding energies (334.8 and 340.2 eV) are in precise agreement with the typical metallic palladium (Pd^0) 3d electron binding energies, indicating that the reductive Pd^0 species was formed in the Pd(dppf)Cl$_2$-catalyzed homocoupling of aromatic halides. [47, 48] The intermediate electron binding energy of Figure 1B may be associated with a mixture of the zero-valent and divalent palladium species.

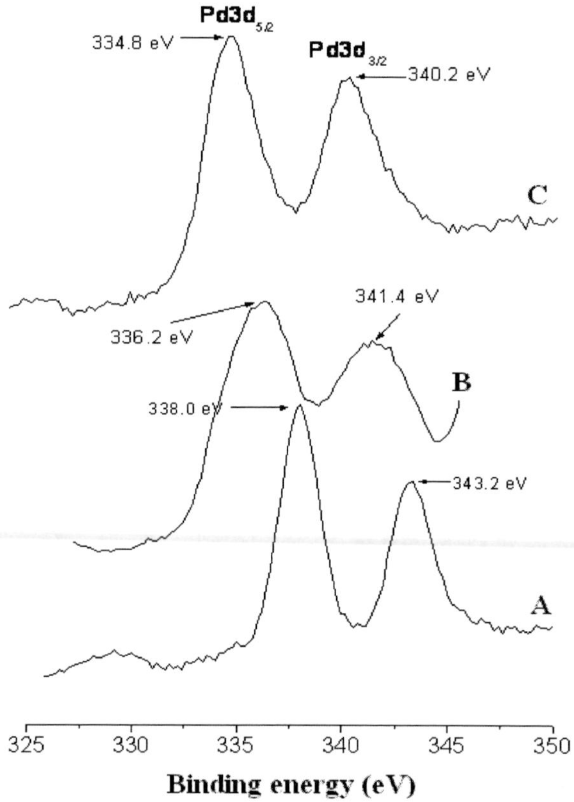

Figure 1. X-ray photoelectron spectra of the palladium catalyst samples. (A) Pd(dppf)Cl$_2$ in 3-pentanol;   (B) Pd(dppf)Cl$_2$ and CsF in 3-pentanol; (C) Pd(dppf)Cl$_2$, CsF and iodobenzene in 3-pentanol.

Most interestingly, the homocoupling biphenyl product is linearly correlated with the ketone (3-pentanone) from the oxidation of the 3-pentanol solvent molecules, indicating that the Pd(dppf)Cl$_2$-catalyzed homocoupling of aromatic halides in alcohols can be characterized by two concomitant palladium mediated redox reactions, i.e. the Pd^0-mediated reduction of aromatic halides and the Pd^{2+}-mediated oxidation of alcohol. [45]

Homogeneous and Heterogeneous Palladium Catalysts. The majority of the palladium catalysis of the aromatic halides has been performed in homogeneous systems. Despite its excellent activities and selectivities, homogeneous catalysis suffers from a number of drawbacks, primarily the difficulty of separating, recovering and reusing the expensive and toxic palladium catalyst and the ligands. In the pharmaceutical and food industries, the resulting palladium contamination in the final products is of major concern because it is frequently reached at an unaccepTable level. [49] In contrast, the heterogeneous catalysts usually do not have these drawbacks, but instead possess many additional attractive features including the ease of handling, product isolation and the straightforward recovery/recycling of the precious metal by filtration, leaving products virtually free of the transition metal residues. [50] Indeed, about 85% of all industrial catalysis processes are based on the heterogeneous catalysts. [51, 52]

Palladium leaching has been shown as a common problem for most heterogeneous palladium catalysts because of the relatively weak chelation of the neutral zero-valent palladium species with the surface molecules of the solid support. [53] Modification of the solid matrix surface structure by chemically binding with special ligand functional groups such as phosphines and *N*-heterocyclic carbenes could enhance the interaction and improve the chelation, which can minimize the palladium leaching. [54] However, the introduction of a surface ligand is typically both challenging and expensive. [55]

Shell-Powder Supported Palladium Heterogeneous Catalysts. Pearl shell, a biogenic hybrid composite, consists of alternating layers of calcium carbonate Tablets separated by thin layers of elastic biopolymers. We have demonstrated that the incorporation of the pearl shell powders could significantly improve the mechanical properties of the epoxy resin composites and attributed the improvement to the intermolecular interactions between the surface chitin and protein molecules of the pearl shell powders with the surrounding epoxy resin matrix molecules. [56] These observations led us to explore the natural pearl shell powders as the solid support to immobilize palladium catalyst. Indeed, the natural pearl shell powder supported palladium can efficiently catalyze various homocoupling [57] and cross-coupling reactions of aromatic halides. [58]

Interestingly, the catalytic activities of the pearl shell powder supported palladium were found unchanged even after five time reuses in the polar DMSO solution. [57] The extremely high catalytic activities and stabilities of the pearl shell powder supported palladium catalysts have been attributed to the strong intermolecular chelation of the zero-valent palladium species with the polar surface hydroxyl, amine and carbonyl functional groups of the natural pearl shell powders. The schematic description of these intermolecular interaction is shown in Figure 2.

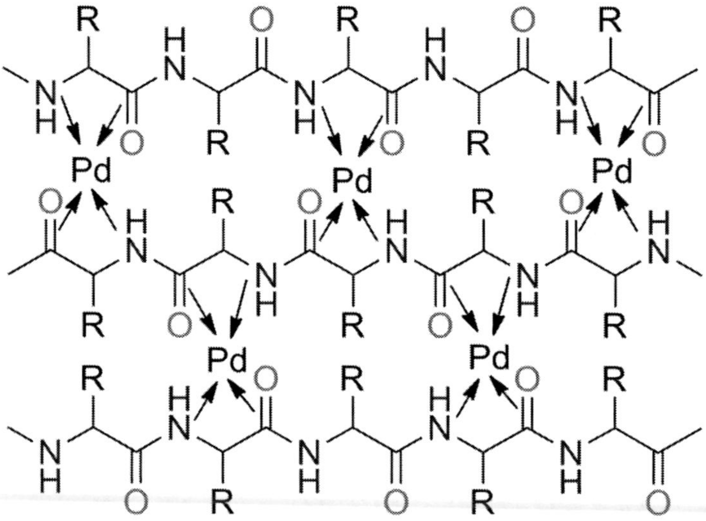

Figure 2. Possible interaction of the palladium species with the surface polar functional groups of the natural shell powders.

Figure 3 presents the X-ray photoelectron spectra of the shell powder supported heterogeneous palladium catalyst. Figure 3A shows that the precatalyst Pd^{2+}/SP has the characteristic palladium $3d_{5/2}$ electron binding energy at 337.9 eV, which is identical to that of the divalent palladium of palladium chloride ($PdCl_2$) (Figure 3D), suggesting that the palladium valence state is conserved onto the surface of the natural pearl shell powders during immobilization. Figures 3B and 3C show the regeneration of the zero-valent palladium species from the reduction of the oxidative Pd^{2+} species during the Pd/SP-catalyzed reductive homocoupling of iodobenzene in ethanol/DMSO solution, which is similar to the *in situ* formation of the reductive palladium species in DMSO [42] and alcohol solution. [57] In other words, the heterogeneous Pd/SP-catalyzed homocoupling of aromatic halides should proceed via the same catalytic mechanism as that proposed for the homogeneous palladium catalysis. [42, 57]

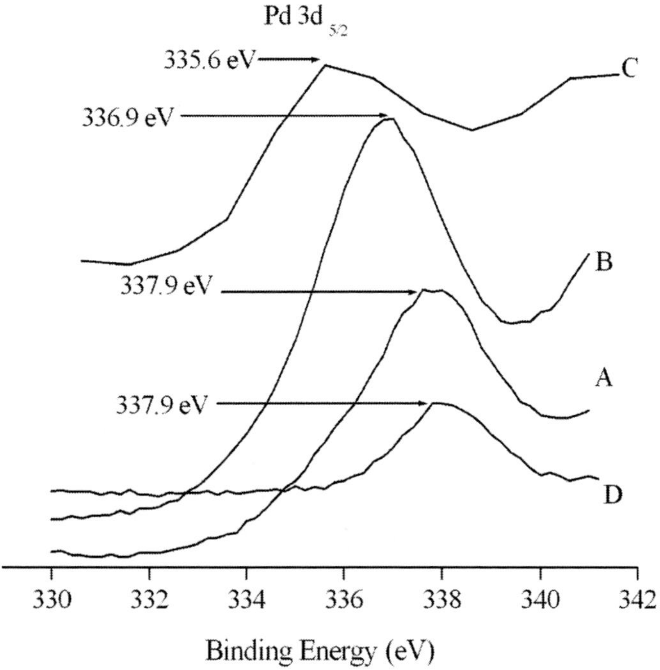

Figure 3. X-ray photoelectron spectra (XPS): (A) Precatalyst Pd2+/SP; (B) Pd/SP catalyst after heated in ethanol/DMSO solution; (C) Pd/SP catalyst isolated from the ethanol/DMSO solution of iodobenzene and potassium acetate; (D) Fresh divalent palladium (Pd2+) species.

Chitosan-based Heterogeneous Palladium Catalysts. Heterogenization of a homogeneous catalysis will dramatically reduce the accessibility of the catalytic centers because the catalytic reactions occur only on the heterogeneous surface, rather than all accessible catalytic centers in the homogeneous system. Therefore, the solid matrices are typically fabricated into fine particles to maximize the surface area in order to enhance the catalytic activities. However, it is a tedious and time-consuming process to separate and recycle the particulate catalysts due to the high pressure drop required for filtration.

The excellent catalytic activities of the crustacean shell powder supported palladium catalyst led us to explore chitosan as the solid support for immobilizing the palladium catalyst. Chitosan is the *N*-deacetylated derivative of chitin, which is the primary organic component of the natural crustacean shell powders and has high affinity for transition metals. [59] Chitosan can easily be prepared into microspheres or flakes with different size and shapes because it is soluble in acidic medium, but essentially insoluble in neutral or basic media. We have prepared the highly porous chitosan microspheres (PCMS) from the chitosan/poly ethylene glycol (PEG) microspheres, followed by removal of the water soluble PEG component. [60] Figure 4 presents the cross-section scanning electron microscopic (SEM) image of the highly porous chitosan microsphere (PCMS). The PCMS supported palladium has been shown to be highly effective catalyst for both homocoupling and cross-coupling reactions of aromatic halides. [60] Some of the representative results are summarized in Table 4 for the Pd/PCMS-catalyzed Heck reactions (Scheme 10) of acrylates with various aromatic iodides.

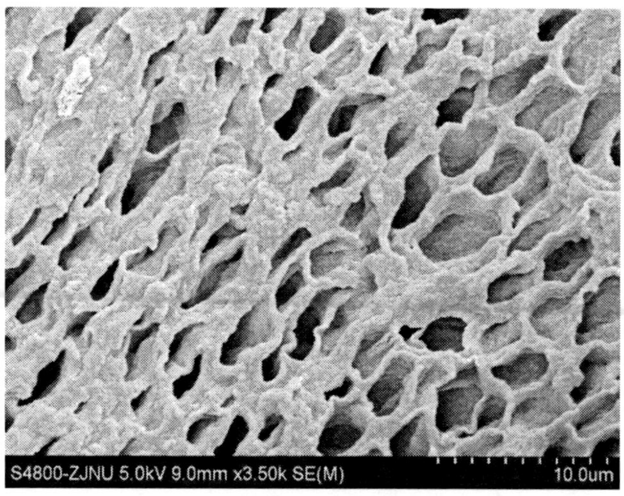

Figure 4. Scanning electron microscopic (SEM) images of the cross-section structure of the palladium coated porous chitosan microspheres (Pd/PCMS).

Most interestingly, the Pd/PCMS catalytic system can also be performed in the environmentally friendly aqueous solution, clearly due to the hydrophilicity of chitosan biomacromolecules. [60] No degradation of the catalytic activities was observed for the heterogeneous Pd/PCMS catalyst even after 15 reuses for the Heck reactions. The Pd/PCMS catalytic activities are comparable to the corresponding homogeneous palladium catalysts. [60] The high stabilities and catalytic activities of the PCMS supported palladium catalytic system can be attributed to the strong interaction of the palladium species with the abundant surface polar functional groups. The porous structure can provide additional surface area to enhance the catalytic activities. These unique properties of the highly porous chitosan microspheres are difficult to obtain if not impossible, by the surface derivatization methods as reported in the literature.

Scheme 10.

Table 4. Pd/PCMS-Catalyzed Heck Cross-coupling of Various Aromatic Iodides with Different Acrylates[a]

Aromatic Halides	Acrylate Substrates	Cross-coupling yield (%)[b]
C_6H_5I	$CH_2{=}CHCO_2(n\text{-}C_4H_9)$	99.7 (*trans*)
$4\text{-}CH_3C_6H_4I$	$CH_2{=}CHCO_2(n\text{-}C_4H_9)$	100 (*trans*)
$4\text{-}FC_6H_4I$	$CH_2{=}CHCO_2(n\text{-}C_4H_9)$	95.9 (*trans*)
$2\text{-}CH_3OC_6H_4I$	$CH_2{=}CHCO_2(n\text{-}C_4H_9)$	83.2 (*trans*)
$4\text{-}BrC_6H_4I$	$CH_2{=}CHCO_2(n\text{-}C_4H_9)$	95.3 (*trans*)
$3\text{-}CH_3OC_6H_4I$	$CH_2{=}CHCO_2(n\text{-}C_4H_9)$	98.5 (*trans*)
$3\text{-}ClC_6H_4I$	$CH_2{=}CHCO_2(n\text{-}C_4H_9)$	100 (*trans*)
$2\text{-}ClC_6H_4I$	$CH_2{=}CHCO_2(n\text{-}C_4H_9)$	97.2 (*trans*)
$3\text{-}FC_6H_4I$	$CH_2{=}CHCO_2(n\text{-}C_4H_9)$	100 (*trans*)
$2\text{-}CH_3C_6H_4I$	$CH_2{=}CHCO_2(n\text{-}C_4H_9)$	100 (*trans*)
$4\text{-}NO_2C_6H_4I$	$CH_2{=}CHCO_2CH_3$	90.8 (*trans*)
C_6H_5I	$CH_2{=}CHCO_2CH_3$	98.4 (*trans*)
$4\text{-}FC_6H_4I$	$CH_2{=}CHCO_2CH_3$	95.2 (*trans*)
$4\text{-}CH_3C_6H_4I$	$CH_2{=}CHCO_2CH_3$	97.8 (*trans*)
C_6H_5I	$CH_2{=}C(CH_3)CO_2CH_3$	54.5 (*trans*), 43.3 (*cis*)
$4\text{-}CH_3C_6H_4I$	$CH_2{=}C(CH_3)CO_2CH_3$	56.4 (*trans*), 42.2 (*cis*)

[a] Reaction conditions: 1.0 mmol aromatic iodide and 2.0 mmol acrylate, 0.01 mmol Pd/PCMS; [b] Cross-coupling product yields.

Poly(vinyl alcohol) (PVA)-based Heterogeneous Palladium Catalysts.
Poly (vinyl alcohol) (PVA), a non-toxic and biodegradable industrial polymer, is an ideal solid support for the immobilization of palladium catalyst because of the plentiful hydroxyl functional groups on its surface. However, it is difficult to fabricate PVA into a suiTable solid matrix with a large surface area due to its semi-crystalline structure and high melting point (230 °C). [61] Thus, the unique electrospinning technique is required to fabricate the PVA porous nanofiber mats in order to mimic the natural pearl shell powders. [62] Figure 5 presents the scanning electron microscopic (SEM) image of the prepared PVA porous nanofiber mats.

PVA porous nanofiber mats were shown as excellent solid support for the immobilization of palladium catalyst and the prepared heterogeneous palladium can efficiently catalyze the Ullmann, Heck-Mizoroki and Sonogashira coupling reactions of aromatic halides. Table 5 presents the results obtained from the Pd/PVA nanofiber mats-catalyzed Sonogashira cross-coupling reactions (Scheme 11).

Scheme 11.

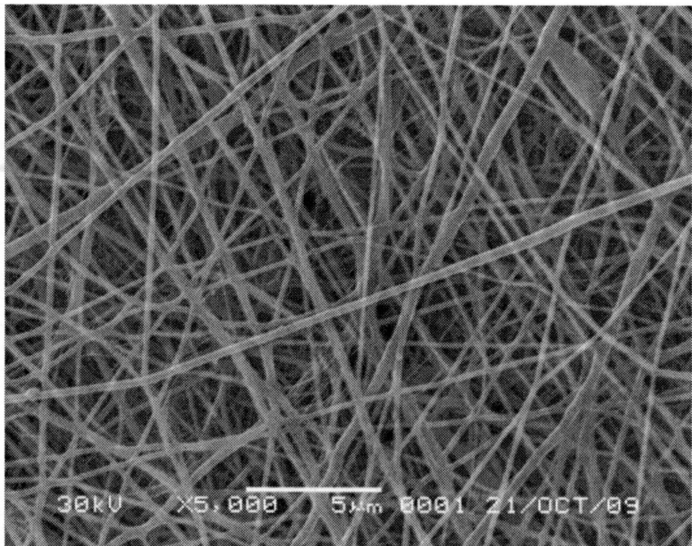

Figure 5. Scanning electron microscopic (SEM) image of the electrospun nanofiber mats.

Table 5. Pd/PVA nanofiber mat-catalyzed Sonogashira cross-coupling reactions of various aromatic iodides with phenylacetylene[a]

Substrate	Yield[b] (%)
C_6H_5I	99.5
$2\text{-}CH_3C_6H_4I$	95.6
$4\text{-}CH_3C_6H_4I$	99.0
$4\text{-}FC_6H_4I$	88.1
$2\text{-}ClC_6H_4I$	88.3

[a] Reaction conditions: 1.0 mmol aromatic halide, 1.2 mmol phenylacetylene, 7.5 mmol potassium acetate and 0.02 mmol Pd/PVA nanofiber mats; [b] Cross-coupling product yields.

Table 6 presents the Heck cross-coupling of iodobenzene with *n*-butyl acrylate as catalyzed by both homogeneous and heterogeneous catalysts. It is worth noting that the PVA nanofiber mat-supported palladium catalyst has comparable catalytic activity with these commonly used homogeneous palladium catalysts, implying that the chelation of palladium species by the abundant PVA surface hydroxyl functional groups may not only improve the palladium catalyst stability, but also enhance its catalytic activities to compensate for the less surface-bound catalytic centers. Moreover, the much larger structure of the nanofiber mats and the porous chitosan microspheres supported catalyst has the advantage of much easier separation and recycling compared to the corresponding particulate catalyst from the reaction media by filtration. [60, 62]

Table 6. Heck cross-coupling reactions of iodobenzene with *n*-butyl acrylate as catalyzed by different palladium catalysts[a]

Palladium catalyst	Yield[b] (%)
$PVA/PdCl_2$ nanofiber mats	93.4
$PdCl_2$	73.4
$Pd(PPh_3)_2Cl_2$	93.2
$Pd(dppf)Cl_2$	99.0

[a] Reaction conditions: 1.0 mmol iodobenzene, 2.0 mmol *n*-butyl acrylate, 3.0 mmol potassium acetate and 0.02 mmol palladium catalyst; [b] Cross-coupling yields.

REFERENCES

[1] Gribble, G.W. Acc. Chem. Res. 1998, 31, 141-152; (b) Gribble, G. W. *Prog. Chem. Org. Nat. Prod.* 1996, 68, 1-423.

[2] Beck, U.; Loser, E. Ullmann's Encyclopedia of Industrial Chemistry, Wiley, Weinheim, 2012, 8, pp. 483-519.

[3] Patai, S.; Rappoport, Z. Halides, Pseudo-Halides and Azides, Wiley, 2004.

[4] Tsuji, J. Palladium reagents and catalysts, Wiley, 2nd Ed., New York, 2004.

[5] (a) Cowdrey, W. A.; Hughes, E. D.; Ingold, C. K.; Masterman, S.; Scott, A. D. *J. Chem. Soc.* 1937, 1252; (b) Ingold, C. K. Structure and Mechanism in Organic Chemistry. 2nd Ed.; Ithaca, New York: Cornell University Press, 1967.

[6] (a) Hale, J. W.; Britton, E. C. Dow Chemical, US Patent 1607618, 1926; (b) Hale, J. W.; Britton, E. C. *Ind. Eng. Chem.* 1928, 20, 114-124.

[7] (a) Williams, W. H. et al. Dow Chemical US Patent, 2432551, 1947; (b) Williams, W. H. et al. Dow Chemical US Patent, 2432552, 1947.

[8] (a) Roberts, J. D.; Simmons, H. E.; Carlsmith, L. A.; Vaughan, C. W. *J. Am. Chem.* Soc. 1953, 75, 3290-3291; (b) Roberts, J. D.; Semenow, D. A.; Simmons, H. E.; Carlsmith, L. A. *J. Am. Chem. Soc.* 1956, 78, 601-611; (c) Roberts, J. D.; Vaughan, C. W.; Carlsmith, L. A.; Semenow, D. A. *J. Am. Chem. Soc.* 1956, 78, 611-614.

[9] Bryce, M. R.; Vernon, J. M. *Adv. Heterocycl. Chem.* 1981, 28, 183-229.

[10] (a) Terrier, F. *Chem. Rev.* 1982, 82, 77-152; (b) Illuminati, G.; Stegel, F. *Adv. Heterocycl. Chem.* 1983, 34, 305-310.

[11] (a) Kim, J. K.; Bunnett, J. F. *J. Am. Chem. Soc.* 1970, 92, 7463; (b) Kim, J. K.; Bunnett, J. F. *J. Am. Chem. Soc.* 1970, 92, 7464.

[12] Bunnett, J. F. *Acc. Chem. Res.* 1978, 11, 413-420.

[13] (a) Kornblum, N.; Michael, R. E.; Kerber, R. C. *J. Am. Chem. Soc.* 1966, 88, 5662; (b) Kornblum, N. *Ang. Chem. Int. Ed.*, 1975, 14, 734-745.

[14] Russell, G. A.; Danen, W. C. *J. Am. Chem. Soc.* 1966, 88, 5663.

[15] (a) Rossi, R. A.; de Rossi, R. H. Aromatic Substitution by the S$_{RN}$1 Mechanism, American Chemical Society, Washington, 1983; (b) Russell, G. A. *Adv. Phys. Org. Chem.* 1987, 23, 271-322; (c) Saveant, J.-M. *Adv. Phys. Org. Chem.* 1990, 26, 130; (d) Rossi, R. A. *Acc. Chem. Res.* 1982, 15, 164-170.

[16] (a) Galli, C. Tetrahedron, 1988, 44, 5205-5208; (b) Katritzky, A. R.; Chen, J. L.; Mareon, C. M.; Maia, A.; Kaehmiri, M. A. Tetrahedron

1986, 42, 101-108; (c) Katritzky, A. R. In Substituent Effects in Radical Chemistry; Viehe, H. G., Janousek, Z., MerBnyi, R., Ma, D. Reidel Publishing Company: Boston, 1986; p 347.

[17]   (a) Yang, D.-L.; Zhang, X.-M.; Cheng, J.-L.; Jai, X.-Q.; Liu, Y.-C. *Acta Chim. Sin.* 1991, 49, 176-182; (b) Zhang, X.-M.; Yang, D.-L.; Liu,Y.-C.; Chen, W.; Chen, W., Cheng, J.-L. *Res. Chem. Intermed.* 1989, 11, 281-300; (c) Zhang, X.-M.; Yang, D.-L.; Liu, Y.-C. *J. Org. Chem.* 1993, 58, 224-227; (d) Zhang, X.-M.; Yang, D.-L.; Jia, X.-Q.; Liu, Y.-C. *J. Org. Chem.* 1993, 58, 7350-7354.

[18]   (a) Zhang, X.-M.; Bordwell, F. G. *J. Org. Chem.* 1992, 57, 4163. (b) Zhang, X.-M.; Bordwell, F. G. *J. Am. Chem. Soc.* 1992, 114, 9787-9792. (c) Zhang, X.-M.; Cheng, J.-P. *Trends in Organic Chemistry*, 1998, 7, 173-179. (d) Luo, R.-L., Guo, Q.-Y; Yu, Q.-S.; Zhang, X.-M. *Chemical Bond Energies in Science Today and Their Applications,* 2008, Hefei, China.

[19]   (a) Bordwell, F. G.; Cheng, J.-P.; Harrelson, J. A. Jr. *J. Am. Chem. Soc.* 1988, 110, 1229-1231; (b) Bordwell, F. G.; Cheng, J.-P.; Ji, G.-Z.; Satish, A. V.; Zhang, X.-M. *J. Am. Chem. Soc.* 1991, 113, 9790-9795.

[20]   (a) Bordwell, F.G.; Zhang, X.-M.; Cheng, J.-P. *J. Org. Chem.* 1993, 58, 6410-6416;

(b) Zhang, X.-M.; Bordwell, F.G. *J. Am. Chem. Soc.* 1994, 116, 904-908.

[21]   Parker, V. D. *Acta Chem. Scand.* 1992, 46, 307-308.

[22]   Zhang, X.-M. *J. Chem. Soc. Perkin Trans.* 2, 1993, 2275-2279.

[23]   Smith, M.; in Augustine, R. L. Reduction Techniques and Applications in Organic Synthesis, Marcel Dekker, NY, 1968, pp. 95-170.

[24]   Heck, R. F. *Acc. Chem. Res.* 1979, 12, 146-151.

[25]   Miller, T. M.; Ahmed, K. J.; Wrighton, M. S. *Inorg. Chem.* 1989, 28, 2347-2355.

[26]   Hassan, J.; Sevignon, M.; Gozzi, C.; Schulz, E.; Lemaire, M. *Chem. Rev.* 2002, 102, 1359-1469.

[27]   (a) Hennings, D. D.; Iwama, T.; Rawal, V. H. *Org. Lett.* 1999, 1, 1205-1208; (b) Kuroboshi, M.; Waki, Y.; Tanaka, H. *J. Org. Chem.* 2003, 68, 3938-3942.

[28]   Stille, J. K.; Lau, K. S. Y. *Acc. Chem Res.* 1977, 10, 434-442.

[29]   (a) Torii, S.; Tanaka, H.; Morisaki, K. *Tetrahedron Lett.* 1985, 26, 1655-1658; (b) Jutand, A.; Negri, S.; Mosleh, A. *J. Chem. Soc., Chem. Commun.* 1992, 1729-1730.

[30] (a) Mukhopadhyay, S.; Rothenberg, G.; Gitis, D.; Wiener, H.; Sasson, Y. *J. Chem. Soc. Perkin Trans.* 2, 1999, 2481-2484; (b) Mukhopadhyay, S.; Rothenberg, G.; Qafisheh, N.; Sasson, Y. *Tetrahedron Lett.* 2001, 42, 6117-6119.

[31] Mukhopadhyay, S.; Rothenberg, G.; Wiener, H.; Sasson, Y. *Tetrahedron*, 1999, 55, 14763-14768.

[32] Brenda, M.; Knebelkamp, A.; Greiner, A.; Heitz, W. *Synlett*, 1991, 809-810.

[33] Hassan, J.; Penalva, V.; Lavenot, L.; Gozzi, C.; Lemaire, M. *Tetrahedron*, 1998, 54, 13793-13804.

[34] Li, J.-H; Xie, Y.-X.; Yin, D.-L. *J. Org. Chem.* 2003, 68, 9867-9869.

[35] Kikukawa, K.; Yamane, T.; Tagaki, M.; Matsuda, T. *J. Chem. Soc. Chem. Commun.* 1972, 695-696.

[36] Grushin, V. V.; Alper, H. *Chem. Rev.* 1994, 94, 1047-1062.

[37] Larock, R. C. Comprehensive Organic Transformations: A Guide to Functional Group Preparations, 2nd Ed., Wiley-VCH, New York, 1999.

[38] Hassan, J.; Sevignon, M.; Gozzi, C.; Schulz, E.; Lemaire, M. *Chem. Rev.* 2002, 102, 1359-1470.

[39] Fanta, P. E. *Chem. Rev.* 1964, 64, 613-632.

[40] Fanta, P. E. *Synthesis* 1974, 9-21.

[41] (a) Jutand, A.; Mosleh, A. *J. Org. Chem.* 1997, 62, 261. (b) Jutand, A.; Negri, S.; Mosleh, A. *J. Chem. Soc. Chem. Commun.* 1992, 1729.

[42] Qi, C.; Sun, X.; Lu, C.; Yang, J.; Du, Y.; Wu, H.; Zhang, X.-M. *J. Organomet. Chem.* 2009, 694, 2912-2916.

[43] Housecroft, C. E.; Owen, S. M.; Raithby, P. R.; Shaykh, B. A. M. *Organometallics*, 1990, 9, 1617-1623.

[44] Zeng, M.; Shao, L.; Qi, Z.; Zhang, X.-M. Ferrocenes: Compounds, Properties and Applications, Phillips, E.S. Ed. Nova Science Publishers, Inc. Chapter 8. 2011.

[45] Zeng, M.; Du, Y.; Shao, L.; Qi, C.; Zhang, X.-M. *J. Org. Chem.* 2010, 75, 2556-2563.

[46] Shao, L.; Du, Y.; Zeng, M.; Li, X.; Shen, W.; Zuo, S.; Lu, Y.; Zhang, X.-M.; Qi, C. *Appl. Organomet. Chem.* 2010, 24, 421-425.

[47] Tsyrul'nikov, P. G.; Afonasenko, T. N.; Koshcheev, S. V.; Boronin, A. I. *Kinet. Catal.* 2007, 48, 728-734.

[48] Kim, K. S.; Gossmann, A. F.; Winograd, N. *Anal. Chem.* 1974, 46, 197-200.

[49] Garrett, C. E.; Prasad, K. *Adv. Synth. Catal.* 2004, 346, 889-900.

[50] (a) Cole-Hamilton, D. J.; Tooze, R. P. Homogeneous Catalysis-Advantages and Problems, in Catalyst Separation, Recovery and Recycling; Chemistry and Process Design; Cole-Hamilton, D. J., Tooze, R. P., Eds.; Springer: Dordrecht, Chapter 1, 2006; (b) Thomas, J. M.; Thomas, W. J. Principles and Practice of Heterogeneous Catalysis; Wiley-VCH: Weinheim, 1996.

[51] Yin, L., Liebscher, *J. Chem. Rev.* 2007, 107, 133-173.

[52] Bhaduri, S.; Mukesh, D. Homogenous Catalysis: Mechanism and Industrial Applications. Wiley-Interscience, New York. 2000. pp: 1~8.

[53] (a) Novak, Z.; Szabo, A.; Repasi, J.; Kotschy, A. *J. Org. Chem.* 2003, 68, 3327-3329; (b) Glasnov, T. N.; Findenig, S.; Kappe, C. O. *Chem. Eur. J.* 2009, 15, 1001-1010; (c) Bhanage, B. M.; Arai, M. *Catal. Rev.* 2001, 43, 315-344.

[54] (a) Gladysz, J. A. *Chem. Rev.* 2002, 102, 3215-3892; (b) Phan, N. T. S.; Van Der Sluys, M.; Jones, C. W. *Adv. Synth. Catal.* 2006, 348, 609-679.

[55] (a) Hoshiya, N.; Isomura, N.; Shimoda, M.; Yoshikawa, H.; Yamashita, Y.; Iizuka, K.; Tsukamoto, S.; Shuto, S.; Arisawa, M. *ChemCatChem.* 2009, 1, 279-285; (b) Crudden, C. M.; Sateesh, M.; Lewis, R. *J. Am. Chem. Soc.* 2005, 127, 10045-10050; (c) Webb, J. D.; MacQuarrie, S.; McEleney, K.; Crudden, C. M. *J. Catal.* 2007, 252, 97-109; (d) Artner, J.; Bautz, H.; Fan, F.; Habicht, W.; Walter, O.; Döring, M.; Arnold, U. *J. Catal.* 2008, 255, 180-189.

[56] (a) Ji, G.; Zhu, H.; Jiang, X.; Qi, C.; Zhang, X.-M. *J. Appl. Polym. Sci.* 2009, 114, 3168-3176; (b) Ji, G.; Zhu, H.; Qi, C.; Zeng, M. P*olymer. Eng. Sci.* 2009, 49, 1383-1388; (c) Sun Zeng, M.; Wang, Y.; Ji, G.; Yao, X.; Shen, Y.; Chen, N.; Yan, F.; Wang, B.; Qi, C. *J. Appl. Polym. Sci.* 2008, 109, 3932-3937.

[57] Zeng, M.; Du, Y.; Qi, C.; Zuo, S.; Li, X.; Shao, L., Zhang, X.-M. *Green Chemistry*, 2011, 13, 350-356.

[58] Shen, Y.-M.; Du, Y.-J.; Zeng, M.-F.; Zhi, D.; Zhao, S.-X.; Rong, L.-M.; Lv, S.-Q.; Du, L.; Qi, C. *Appl. Organometal. Chem.* 2010, 24, 631-635.

[59] Vincent T.; Guibal, E. *Ind. Eng. Chem. Res.* 2002, 41, 5158-5164.

[60] Zeng, M.; Zhang, X.; Du, Y.; Shao, L.; Qi, C.; Zhang, X.-M. *J. Organomet. Chem.* 2012, 704, 29-37.

[61] Rampino, L. D.; Nord, F. F. *J. Am. Chem. Soc.* 1943, 65, 429-431.

[62] (a) Shao, L.; Ji, W.; Dong, P.; Zeng, M.; Qi, C. Zhang, X.-M. *Appl. Catal. A. Gen*, 2012, 413-414, 267-272; (b) Shao, L.; Liu, J.; Ye, Y.; Zhang, X.-M.; Qi. C. *Appl. Organometal. Chem.* 2011, 25, 699-703.

In: Halides ISBN: 978-1-62417-947-1
Editor: Jean Lefebure © 2013 Nova Science Publishers, Inc.

Chapter 4

DISLOCATION PINNED BY A MONOVALENT ION IN VARIOUS ALKALI HALIDE CRYSTALS

Y. Kohzuki[*]

Oshima National College of Maritime Technology, Oshima-gun,
Yamaguchi, Japan

PART I. VARIOUS MODELS BETWEEN A DISLOCATION AND THE MONOVALENT CATION

ABSTRACT

The force-distance ($F(x)$) relation, which represents the model overcoming a weak obstacle by a dislocation with the help of thermal fluctuations, is described here for short-range obstacle less than about 10 atomic diameters. Furthermore, the three models (the square, the parabolic, and the triangular $F(x)$) taking account of the Friedel relation are investigated with respect to the relation between effective stress and temperature for $KCl:Li^+$ and $KCl:Na^+$ single crystals. On the three force-distance relations, the value of T_c is almost the same irrespective of the kind of force-distance relation for both the specimens, and the value of T_c ($254\sim257$ K) for $KCl:Li^+$ is obviously small in comparison with that ($266\sim271$ K) for $KCl:Na^+$. T_c is the critical temperature at which the effective stress becomes zero and a dislocation breaks away from the

---

[*] Corresponding author's email: kouzuki@oshima-k.ac.jp

defects of cubic symmetry around Li^+ or Na^+ only with the help of thermal activation.

MODEL OVERCOMING THE THERMAL OBSTACLE BY A DISLOCATION

Alkali halide crystals have some advantages [1]: (1) simple crystal structure (i.e. rock salt structure), (2) low dislocation density in grown crystal (e.g., 10^4 cm^{-2} order for NaCl [2] and KCl [3] single crystals) as against that of annealed metals (around 10^7 cm^{-2} [4], e.g., 1 to 5×10^6 cm^{-2} in pure Cu single crystals [5], 1 to 5×10^6 cm^{-2} in α -brass single crystals[6], and 10^7 cm^{-2} order for pure Fe single crystals [7]), (3) transparency, and further (4) readily availability for pure single crystal, etc. Pure or impurity-doped alkali halide crystals, therefore, are excellent materials for an investigation on strength of crystals.

A dislocation will encounter a stress field illustrated schematically in figure 1 as it moves through on the slip plane containing many weak obstacles and a few strong ones. In the figure, the positive stress concerning axis of the ordinate opposes the flow stress (applied stress), τ, and the negative stress assists it. Extrinsic resistance to the dislocation motion has two types: one is long-range obstacle (the order of 10 atomic diameters or greater) and the other short-range obstacle (less than about 10 atomic diameters). The former is considered to be forest dislocations, large precipitates or second-phase particles, and grain boundary, for instance, and the latter impurity atoms, isolated and clustered point defects, small precipitates, intersecting dislocations, etc. Overcoming the latter type of obstacles (byname, thermal obstacles) by a dislocation, thermal fluctuations play an important role in aid of the flow stress above the temperature of 0 K. Then the aid energy, ΔG, supplied by the thermal fluctuations is given by the shaded part in figure 1. Thus the dislocation can move through below τ_0. τ_0 is the value of τ at 0 K. As for the long-range obstacles (byname, athermal obstacles), the energy barrier is so large that the thermal fluctuations play no role in overcoming them within the temperature range.

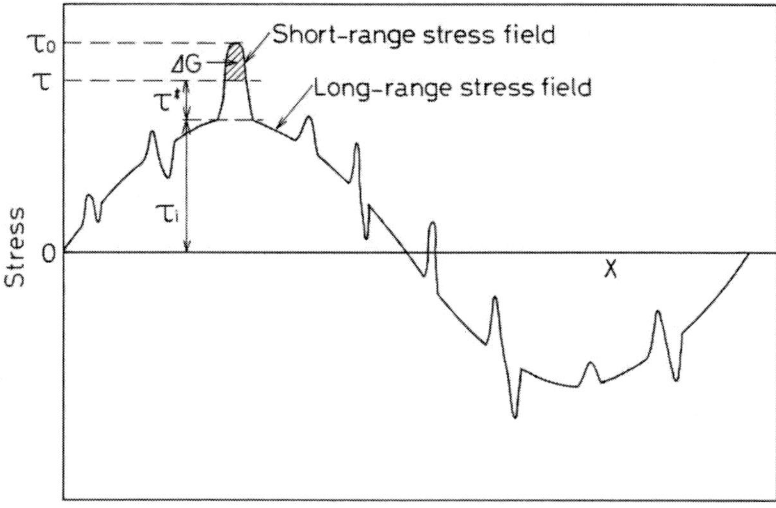

Figure 1. Stress fields encountered by a dislocation moving through the crystal lattice.

The representation of figure 2 is concerned with a common type of thermal activation barrier. The free energy (G) varies with the distance (x) between a dislocation and the obstacle as given in figure 2 (a). When a dislocation overcomes the short-range obstacles, the free energy becomes high on account of the work (ΔW) done by the applied stress. Then the resistance (F), where it can be defined by the differentiation of free energy with respect to x (i.e. $\partial G/\partial x$), to the dislocation motion is revealed as figure 2 (b) in accord with the abscissa of figure 2 (a). Figure 2 (b) corresponds to typical force-distance curve for short-range obstacle among those in figure 1. Shape of this curve represented by $F(x)$ means the model overcoming the obstacle by a dislocation. ΔG_0, which is taken as the shaded area under $F(x)$ in figure 2 (b), is the Gibbs free energy of activation for the breakaway of the dislocation from the obstacle in the absence of an applied stress (in this case it is equivalent to the Helmholtz free energy for the dislocation motion).

The separation of various models between a dislocation and the obstacle for a particular case has been conducted from two methods (i.e., τ_0^* versus c and τ^* versus T) [8-11] as mentioned below. The models of Mott-Nabarro, Seeger, and Friedel lead to $\tau_0^* \propto c$ and the Fleischer's model $\tau_0^* \propto c^{1/2}$, provided $c > 10^{-4}$. τ_0^* is the thermal component of flow stress at 0 K and c the concentration of point defects. With respect to the effective stress(τ^*)-temperature(T) relationship, the Mott-Nabarro model yields a straight line in the τ^*–$T^{2/3}$ plot, the Seeger's model in the $\tau^{*2/3}$–$T^{2/3}$ plot, the Friedel's model in the τ^*–T plot, the Fleischer's model in the $\tau^{*1/2}$–$T^{1/2}$ plot, the modified Fleischer's model in the $\tau^{*1/3}$–$T^{1/2}$ plot, the square potential in the $\tau^{*2/3}$–T plot, the parabolic potential in the $\tau^{*2/3}$–$T^{2/3}$ plot, and the triangular potential in the $\tau^{*2/3}$–$T^{1/2}$ plot.

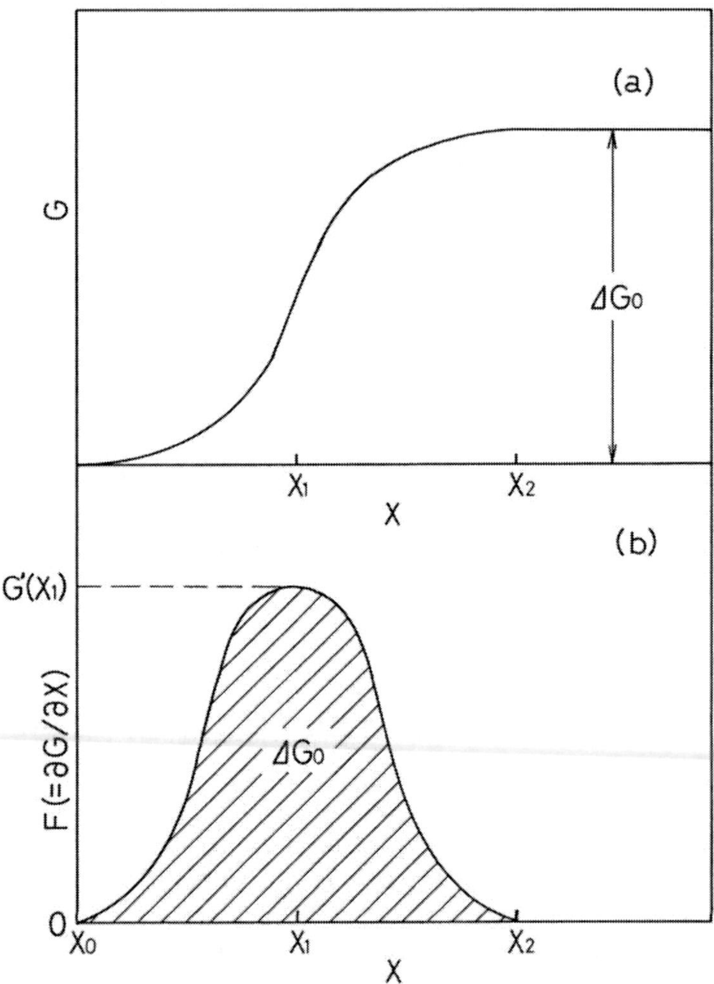

Figure 2. The process for thermal activated overcoming of the short-range obstacle by a dislocation. Variation in (a) the Gibbs free energy of activation and (b) the force acted on the dislocation with the distance for a dislocation motion.

EFFECTIVE STRESS DUE TO IMPURITIES VS. TEMPERATURE

Several models for the force-distance profile and the relation between temperature and the effective stress due to the obstacles have been proposed [12, 13]. For example, the relation between temperature and the stress is as follows

$$\left(\frac{\tau^*}{\tau_0^*}\right)^{2/3} = 1 - \left(\frac{T}{T_c}\right)^{p/(p+1)},$$

(1)

where T_c is the critical temperature at which the effective stress becomes zero and a dislocation breaks away from the impurities only with the help of thermal activation, and p is a parameter. The force-distance curve is triangular (TR) when p is one [14], parabolic (PA) when p is two, and square (SQ) when p is infinite [13], where the Friedel relation [15] between the effective stress and the average length of dislocation segments is taken into account. Determination of T_c has been attempted from the temperature dependence of the yield stress (τ_y) so far [16–22], because τ_y appears to approach an almost constant value above T_c. Then, the effective stress due to impurities was defined by subtraction of the almost constant yield stress at high temperature (such as room temperature) from the yield stress at a given low temperature. That is, the yield stress at high temperature was taken as the internal stress (τ_i) for the specimen (see figure 3). However, the value of internal stress in this assumption may be doubtful, since the yield stress seems to be gradually decreased over the temperature of T_c. This is because an athermal component of yield stress depends on temperature only through the shear modulus. Further, it is difficult to investigate the interaction between a dislocation and weak obstacles during plastic deformation from the yield stress for the reason that yield stress depends on dislocation velocity, dislocation density and multiplication of dislocations [22]. On the other hand, the value of τ_{pl}, which can be obtained from the first bending point of the strain-rate sensitivity versus stress decrement curve at a given strain, has been reported to be depend on temperature, and on the type and the density of weak obstacle [23]. Observation of τ_{pl}, therefore, has provided information on the interaction between a dislocation and weak obstacles. That is, the temperature dependence of τ_{pl} reveals the force-distance profile, which expresses the interaction between a dislocation and a weak obstacle. When alkali halide crystals are doped with monovalent ions, cubic symmetry defects are produced around the ions. Then, assuming that the interaction between a dislocation and the point defect can be approximated to TR, PA or SQ, the linear relationship between the effective stress and temperature is shown in figure 4 (a) to (c) for KCl:Li$^+$ (0.5 mol. % in the melt) and figure 5 (a) to (c) for KCl:Na$^+$ (0.5 mol. % in the melt). The slopes of straight lines in figures 4 (a) to (c) and 5 (a) to (c) were determined by the method of least squares. The temperatures of T_c, at which the lines intersect the abscissa in these figures and τ_{p1} is zero, are determined to be 254 to 257 K for KCl:Li$^+$ and 266 to 271 K for KCl:Na$^+$ as given in Table 1. On the three force-distance relations, the value of T_c is almost the same irrespective of the kind of force-distance relation for both the specimens,

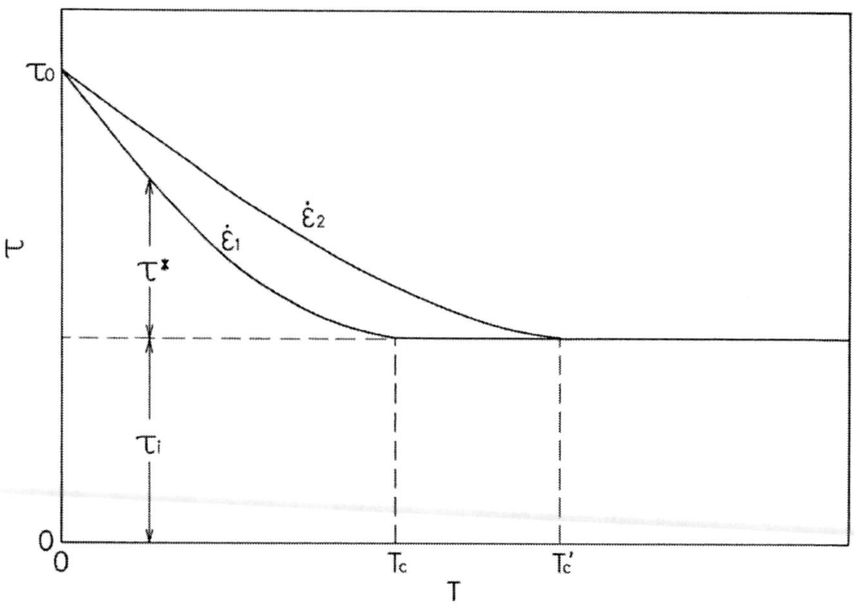

Figure 3. Variation of flow stress with temperature at two strain rates ($\dot{\varepsilon}_1 < \dot{\varepsilon}_2$).

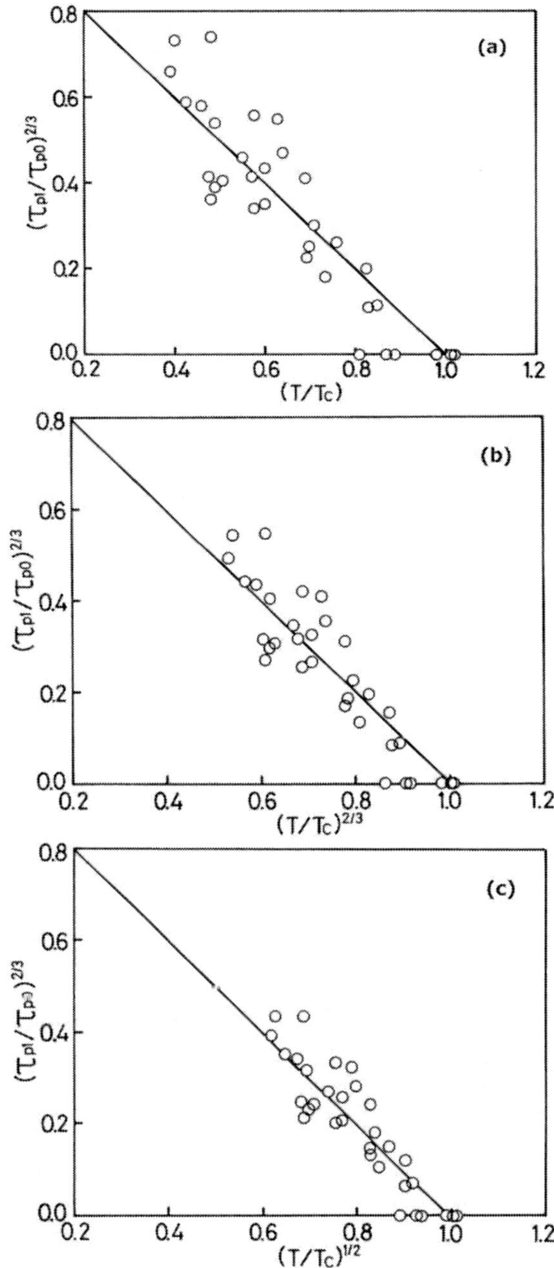

Figure 4. Relationship between the effective stress and the temperature for KCl:Li$^+$ (0.5 mol.% in the melt) at various models: (a) the square, (b) the parabolic, and (c) the triangular force-distance relations.

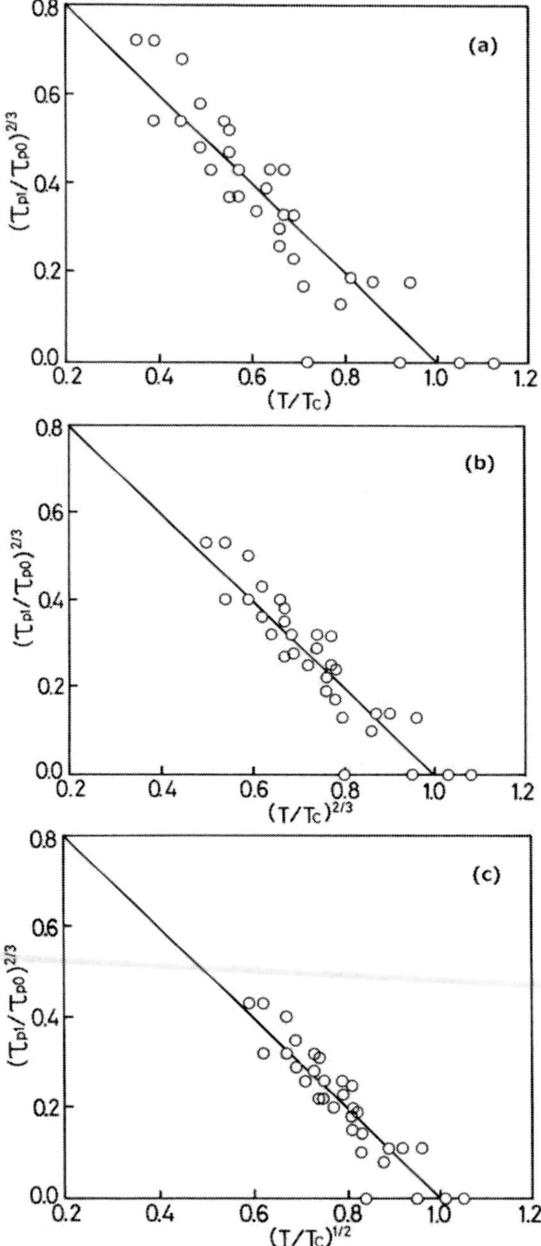

Figure 5. Relationship between the effective stress and the temperature for KCl:Na$^+$ (0.5 mol.% in the melt) at various models: (a) the square, (b) the parabolic, and (c) the triangular force-distance relations.

Table 1. Values of T_c and τ_{p0} for the three force-distance relations between a dislocation and the monovalent cation

The force-distance relation	Specimen[*]	T_c (K)	τ_{p0} (MPa)
Square	KCl:Li$^+$	254	1.10
	KCl:Na$^+$	267	1.45
Parabola	KCl:Li$^+$	256	1.73
	KCl:Na$^+$	266	2.29
Triangle	KCl:Li$^+$	257	2.44
	KCl:Na$^+$	271	3.16

[*] The concentration of Li$^+$ or Na$^+$ in KCl is 0.50 mol.% in the melt.

and the value of T_c for KCl:Li$^+$ is obviously small in comparison with that for KCl:Na$^+$ from Table 1. At the temperature of T_c, thermal fluctuations can provide the entire energy for breaking through the impurity. T_c is not influenced by the concentration of impurities (i.e. Li$^+$ or Na$^+$ here) [23-26], whereas it depends on strain rate in accordance with

$$T_c = \left(\Delta G_0/k\right)\ln\left[\dot{\varepsilon}/\left\{\rho b^2 v_D \left(L_0/L\right)^2\right\}\right], \tag{2}$$

where ΔG_0 is the Gibbs free energy for the breakaway of a dislocation from an impurity, k the Boltzmann constant, ρ the density of mobile dislocations, b the magnitude of the Burgers vector, and v_D the Debye frequency.

REFERENCES

[1]   Kataoka, T. *Doctoral Thesis at Osaka Univ.;* Osaka, 1975; p 2 (in Japanese).

[2]   Argon, A. S.; Nigam, A. K.; Padawer, G. E. *Phil. Mag.* 1972, *25*, 1095-1118.

[3]   Kohzuki, Y. *Doctoral Thesis at Kanazawa Univ.;* Kanazawa, 1994, pp 12-13.

[4]   Suzuki, H. *Introduction to Theory of Dislocations*; AGNE; Tokyo, 1989; p 70 (in Japanese).

[5]   Young, Jr., F. W. *J. Appl. Phys.* 1961, *32*, 192-201.

[6]   Meakin, J. D.; Wilsdorf, H. G. F. *Trans. Met. Soc. AIME* 1960, *218*, 737-745.

[7]   Takeuchi, S. *J. Phys. Soc. Jap.* 1969, *27*, 929-940.

[8]   Conrad, H. *Acta Metall.* 1966, *14*, 1631-1633.

[9]   Conrad, H. *Can. J. Phys.* 1967, *45*, 581-590.

[10]   Sprackling, M. T. The Plastic Deformation of Simple Ionic Crystals; (edited by Alper, A. M.; Margrave, J. L.; Nowick, A. S.) Academic Press; London, 1976; p 188.

[11]   Koppenaal, T. J.; FINE, M. E. *Trans. Metall. Soc. AIME* 1962, *224*, 347-353.

[12]   Cottrell, A. H.; Bilby, B. A. *Proc. Phys. Soc. A* 1949, *62*, 49-62.

[13]   Seeger, A. *Phil. Mag.* 1955, *46*, 1194-1217.

[14]   Foreman, A. J. E.; Makin, M. J. *Phil. Mag.* 1966, *14*, 911-924.

[15]   Friedel, J. *Dislocations*; Pergamon Press; Oxford, 1964; p 224.

[16]   Suszyńska, M. *Kristall. Technik* 1974, *9*, 1199-1207.

[17]   Dryden, J. S.; Morimoto, S.; Cook, J. S. *Phil. Mag.* 1965, *12*, 379-391.

[18]   Stoloff, N. S.; Lezius, D. K.; Johnston, T. L. *J. Appl. Phys.* 1963, *34*, 3315-3322.

[19]   Chin, G. Y.; Van Uitert, L. G.; Green, M. L.; Zydzik, G. J.; Kometani, T. Y. *J. Am. Ceram. Soc.* 1973, *56*, 369-372.

[20]   Kataoka, T.; Yamada, T. *Jpn. J. Appl. Phys.* 1977, *16*, 1119-1126.

[21]   Guiu, F.; Langdon, T. G. *Phil. Mag.* 1974, *30*, 145-160.

[22]   Johnston, W. G.; Gilman, J. J. *J. Appl. Phys.* 1959, *30*, 129-144.

[23]   Kohzuki, Y.; Ohgaku, T.; Takeuchi, N. *J. Mater. Sci.* 1993, *28*, 3612-3616.

[24]   Hikata, A.; Johnson, R. A.; Elbaum, C. *Phys. Rev. B* 1970, *2*, 4856-4863.

[25]   Ohgaku, T.; Takeuchi, N. *Phys. Status Solidi (a)* 1989, *111*, 165-172.

[26]   Ohgaku, T.; Takeuchi, N. *Phys. Status Solidi (a)* 1992, *134*, 397-404.

PART II. CHARACTERISTICS ON THE BREAKAWAY OF A DISLOCATION FROM THE MONOVALENT CATION

ABSTRACT

The energy for the breakaway of a dislocation from the monovalent cation is obtained through the slope of linear relationship of reciprocal of temperature ($1/T$) versus $\tau_{p1}(\Delta \ln \dot{\varepsilon} / \Delta \tau')_{p}$. Here τ_{p1} represents the effective stress and $(\Delta \ln \dot{\varepsilon} / \Delta \tau')_{p}$ the reciprocal of strain-rate sensitivity due to the additive cations when a dislocation moves forward with the help of oscillation during plastic deformation. Analyzing this result, it would be deduced that the breaking angle ϕ_0 is around $178°$ and $(L/L_0) \approx 7.0$ for KCl: Li$^+$ or Na$^+$ single crystals. L is the average length of dislocation segments and L_0 the average spacing of monovalent cations on a slip plane.

INTRODUCTION

Solution hardening may be divided into two classes in terms of the different types of defects around additive ions. That is to say, the rapid hardening is due to the asymmetrical distortions in ionic crystals doped with aliovalent cations and the gradual hardening due to the defects of cubic symmetry around isovalent cations in ionic crystals [1-3]. Roughly speaking, the value of $\Delta \tau / \Delta c^{1/2}$ (an increase in flow stress per the square root of concentration of point defects) for the rapid hardening is over several tens times as large as that for the gradual hardening in alkali halide crystals contained the point defects at low concentration below 10^{-4}.

The preceding part I of this series dealt with the force-distance curve for short-range obstacle to the dislocation motion and the critical temperature, at which the effective stress becomes zero and a dislocation breaks away from the defect of cubic symmetry around the monovalent cation only with the help of thermal activation. In part II of this series, the Gibbs free energy of activation overcoming of the short-range obstacle by a dislocation is evaluated by analyzing the data derived from the strain-rate cycling tests associated with oscillation for KCl: Li$^+$ (0.5 mol.% in the melt) or Na$^+$ (0.5 mol.% in the melt) single crystals.

BREAKING ANGLE AND (L/L_0) FOR KCl:Li$^+$ OR Na$^+$ SINGLE CRYSTALS

When the thermally activated overcoming of the substitutional impurities controls the dislocation velocity [4], the strain rate is assumed to be given by an Arrhenius-type equation [5]:

$$\dot{\varepsilon} = \dot{\varepsilon}_0 \exp\left(\frac{-\Delta G}{kT}\right),$$

(1)

where $\dot{\varepsilon}_0$ is a frequency factor, ΔG the change in Gibbs free energy of activation for the dislocation motion, and kT has usual meaning. The Gibbs free energy for the square force-distance profile (SQ) between a dislocation and the impurity is given by [6]

$$\Delta G = \Delta G_0 - \beta \tau^{*2/3} \quad (\beta = (2\mu b^4 d^3 L_0^2)^{1/3}),$$

(2)

where μ is the shear modulus, b the magnitude of the Burgers vector, d the activation distance, and L_0 the average spacing of obstacles distributed randomly on a slip plane. Substituting equation (2) into equation (1) and differentiating with respect to the effective stress (τ^*), we find

$$\tau^* \frac{\partial \ln \dot{\varepsilon}}{\partial \tau^*} = \frac{2\Delta G_0}{3kT} + \frac{2}{3}\ln(\dot{\varepsilon}/\dot{\varepsilon}_0)$$

(3)

The Gibbs free energy (ΔG_0) for the breakaway of a dislocation from the defect around additive cation (see figure 2 in preceding part I of this series), which is obtained through the slope of straight line of $\tau^*(\partial \ln \dot{\varepsilon}/\partial \tau^*)$ versus $(1/T)$ in equation (3), is 0.73 eV for KCl:Li$^+$ and 0.62 eV for KCl:Na$^+$. Here, the value of $\tau^*(\partial \ln \dot{\varepsilon}/\partial \tau^*)$ was evaluated by $\tau_{p1}(\Delta \ln \dot{\varepsilon}/\Delta \tau')_p$ on the basis of the relative curve between the strain-rate sensitivity $(\Delta \tau'/\Delta \ln \dot{\varepsilon})$ and stress decrement $(\Delta \tau)$ due to oscillation as shown in figure 1 (see detailed article [7]).

By using μ and ΔG_0 in Table 1, it is possible to obtain the breaking angle ϕ_0 (see figure 1 in part III of this series) from the equation [8]:

$$\frac{\phi_0}{2} = \cos^{-1}\left(\frac{\Delta G_0}{2\mu b^3}\right)$$

(4)

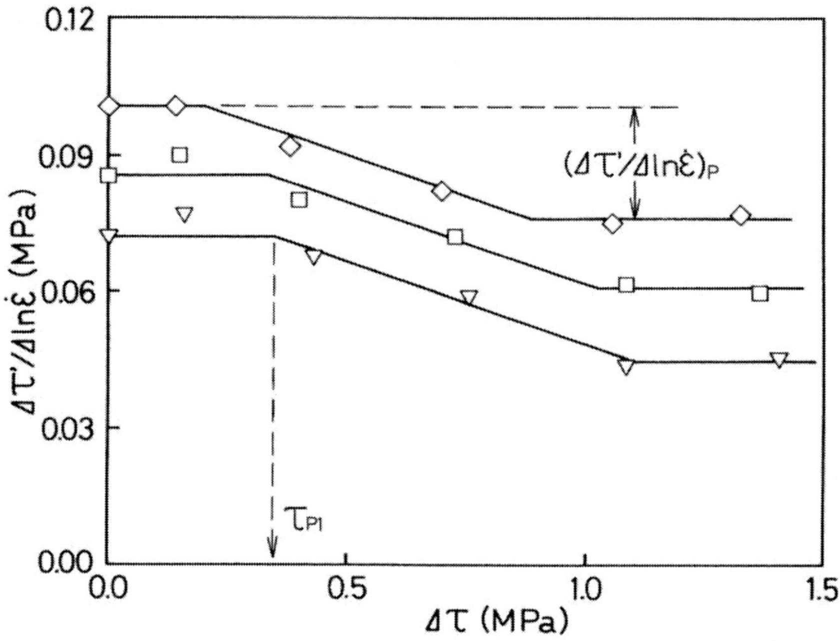

Figure 1. Relation between the strain-rate sensitivity $(\Delta\tau'/\Delta\ln\dot{\varepsilon})$ and the stress decrement $(\Delta\tau)$ due to oscillation for KCl:Na$^+$ (0.50 mol.% in the melt) at 180 K and strains: $\varepsilon = (\triangledown)$ 17 %, ($\square$) 20 %, and ($\lozenge$) 23 %.

On the other hand, the value of (L/L_0) can also be found on the basis of the following equation [8]:

$$\left(\frac{L_0}{L}\right)^2 = \cos\left(\frac{\phi_0}{2}\right) \tag{5}$$

As shown in Table 1, the calculations for both the crystals are within $177° < \phi_0 < 178°$ and around $6.59 \le (L/L_0) \le 7.07$ at the temperature of 0 K. These values are almost the same, irrespective of the different size of additive cations (i.e. Li$^+$ size: 1.18 $\overset{\circ}{\text{A}}$ and Na$^+$ size: 2.04 $\overset{\circ}{\text{A}}$).

Table 1. Breaking angle ϕ_0 (degrees) and $(L_0/L)^2$ derived from equations (4), (5) and input parameters (the shear modulus μ ($\times 10^{10}$ Pa) for [110] direction, the magnitude of the Burgers vector b (Å) ($=\sqrt{2}r$) at 0 K and the Gibbs free energy ΔG_0 (eV) for the dislocation motion) for the alkali halides

Single crystal	ϕ_0	$(L_0/L)^2$	(L/L_0)	μ^*	b	ΔG_0
KCl:Li$^+$ (0.5 mol.%)	177.4	0.023	6.59	1.01	4.407	0.73 [9]
KCl:Na$^+$ (0.5 mol.%)	177.7	0.020	7.07	1.01	4.407	0.62 [9]

* The values of shear modulus are approximately estimated as the matrix of crystal at about 0 K.

CONCLUSION

It was found that the breaking angle ϕ_0 of the defect of cubic symmetry around monovalent cations (Li$^+$ or Na$^+$) is within 177 to 178 degrees for a mobile dislocation in KCl:Li$^+$ or Na$^+$ single crystals at the temperature of 0 K. The additive cations are the weak obstacles such as impede the mobile dislocation at $\phi_0 \approx 180$ degrees. Furthermore, the values of (L/L_0) were found to be within 6.59 to 7.07 for the crystals. These values are almost the same irrespectively the size of additive cation.

REFERENCES

[1]   Fleischer, R. L. *Acta Metall.* 1962, *10*, 835-842.

[2]   Fleischer, R. L.; Hibbard, W. R. *The Relation between Structure and Mechanical Properties of Metal*; Her Majesty's Stationary Office; London, 1963, p 261.

[3]   Johnston, W. G.; Nadeau, J. S.; Fleischer, R. L. *J. Phys. Soc. Jap.* 1963, *Suppl.* I *18*, 7-15.

[4]   Guiu, F.; Langdon, T. G. *Phil. Mag.* 1974, *30*, 145-160.

[5]   Christian, J. W.; Masters, B. C. *Proc. R. Soc. A* 1964, *281*, 240-257.

[6]   Sumino, K. *Jpn. Inst. Met.* 1972, *11*, 31-47 (in Japanese).

[7]   Kohzuki, Y. *Plasticity of Crystals in a microscopic viewpoint on the basis of dislocation motion*; Nova Science Publishers; New York, 2012.

[8]   Kohzuki, Y. *Dislocation pinned by a monovalent ion in various alkali halide crystals Part III. Breaking angle of an anion as obstacle for moving dislocation* in: *Halides: Chemistry, Physical Properties and Structural Effects*; Nova Science Publishers; New York, 2013.

[9]   Kohzuki, Y. *Phil. Mag.* 2010, *90*, 2273-2287.

PART III. BREAKING ANGLE OF AN ANION AS OBSTACLE FOR MOVING DISLOCATION

ABSTRACT

The breaking angle ϕ_0, at which a dislocation breaks away from a weak obstacle at the temperature of 0 K, measures the strength of the obstacle-dislocation interaction. The ϕ_0 ranges 173° to 177° for the various alkali halides single crystals (NaCl:Br⁻, NaBr: Cl⁻ or I⁻, KCl: Br⁻ or I⁻, and RbCl: Br⁻ or I⁻). The values of (L/L_0) were found to be within 4.05 to 5.87 there. L is the average length of dislocation segments and L_0 the average spacing of monovalent anions on a slip plane.

INTRODUCTION

In the preceding part II of this series, hardening (especially, breaking angle) due to substitutional additive cations (i.e. Li^+ or Na^+ ions), which corresponds to the gradual hardening, was described for two kinds of alkali halide crystals doped with monovalent cations (KCl: Li^+ (0.5 mol.%) or Na^+ (0.5 mol.%)) at the temperature of 0 K. The breaking angle ϕ_0, at which a dislocation breaks away from a weak obstacle at the temperature of 0 K, measures the strength of the obstacle-dislocation interaction.

When a mobile dislocation is pressed against weak rigid obstacle by an effective stress (τ^*), the dislocation is bent by the obstacle on a slip plane as shown in figure 1.

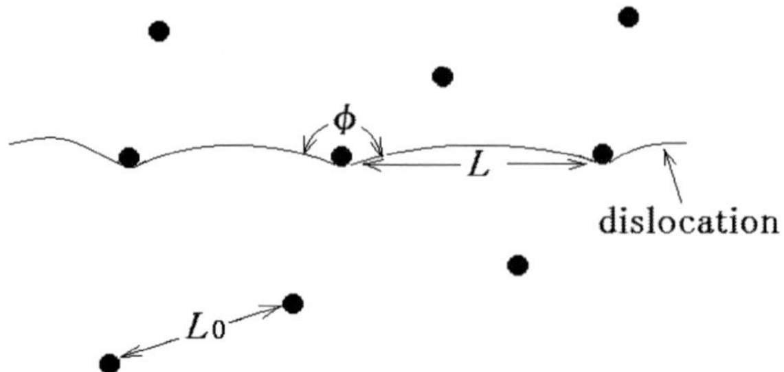

Figure 1. Dislocation held up at weak obstacles on the slip plane. Solid circles (•) represent one type of weak obstacles, such as impurities.

The breaking angle of dislocation by a monovalent anion is herein investigated for various alkali halide single crystals (NaCl:Br⁻, NaBr: Cl⁻ or I⁻, KCl: Br⁻ or I⁻, and RbCl: Br⁻ or I⁻) at the temperature of 0 K. Then, the average spacing (L) of obstacles along the dislocation is approximately represented by [1]

$$L = \left(\frac{2L_0{}^2 E}{\tau^* b} \right)^{1/3} , \tag{1}$$

where L_0 is the average spacing of obstacles distributed randomly on a slip plane, E the line tension of the dislocations, and b the magnitude of the Burgers vector. Equation (1) is termed the Friedel relation and has been applied to most weak obstacles to dislocation motion at low solute concentration. The value of (L/L_0), which is characteristic for each crystal although the parameters L and L_0 depend on the concentration of additive ions, is also presented for the halide crystals.

BREAKING ANGLE AND (L/L_0) FOR THE DISLOCATION MOTION

The interaction between a dislocation and the monovalent anions is investigated for various alkali halide crystals with NaCl-type of structure. This is examined from the breaking angle ϕ_0. The values of ϕ_0 under the effective stress is given by [2]

$$\cos\frac{\phi_0}{2} = \left(\frac{\tau^* L_0 b}{2E}\right)^{2/3} \tag{2}$$

Equation (1) will be transformed to

$$L_0 = \left(\frac{\tau^* L^3 b}{2E}\right)^{1/2} \tag{3}$$

Substituting of equation (3) in equation (2) gives

$$\cos\frac{\phi_0}{2} = \frac{\tau^* Lb}{2E}, \tag{4}$$

that is to say,

$$\frac{\phi_0}{2} = \cos^{-1}\left(\frac{\Delta G_0}{2\mu b^3}\right), \tag{5}$$

where ΔG_0 is the activation energy for overcoming of the defect due to an additive anion by a dislocation and can be represented by $\tau^* L b^2$ and μ the shear modulus. The line tension of the dislocations is replaced by μb^2. The reciprocal of μ is given by [3]

$$\frac{1}{\mu} = S_{44} + 4\left(S_{11} - S_{12} - \frac{1}{2}S_{44}\right)\left(l_1^2 l_2^2 + l_2^2 l_3^2 + l_3^2 l_1^2\right), \tag{6}$$

where S_{11}, S_{12} and S_{44} are the adiabatic elastic compliances and l_1, l_2 and l_3 are the direction cosine of stress-axis. The values $l_1 = 1/\sqrt{2}$, $l_2 = 1/\sqrt{2}$ and $l_3 = 0$ are used for the [110] direction. By using the values of μ derived from equation (6) and ΔG_0 in Table 1, it is possible to obtain the breaking angle ϕ_0 from equation (5), whereupon the magnitude of the Burgers vector of a dislocation with $b = r$ [110] is $\sqrt{2}r$ (i.e., $b = \sqrt{2}r$ whereof r is the interionic distance at 0 K [4]).

Table 1. Bending angle ϕ_0 (degrees) and $(L_0/L)^2$ derived from equations (5), (7) and input parameters (the shear modulus μ ($\times 10^{10}$ Pa) for [110] direction, the magnitude of the Burgers vector b ($\mathring{A}$) ($=\sqrt{2}r$) at 0 K and the Gibbs free energy ΔG_0 (eV) for the dislocation motion) for the NaCl-type alkali halides

Single crystal	ϕ_0	$(L_0/L)^2$	(L/L_0)	μ^*	b	ΔG_0
NaCl:Br$^-$ (0.1, 0.5, 1 mol.%)	176.7	0.029	5.87	1.66	3.944	0.37 [5]
NaBr:Cl$^-$ (1.0 mol.%)	176.3	0.032	5.59	1.371	4.178	0.40 [6]
NaBr:I$^-$ (1.0 mol.%)	175.7	0.038	5.15	1.371	4.178	0.47 [6]
KCl:Br$^-$ (1, 2 mol.%)	174.7~175.8	0.037~0.046	4.65~5.19	1.01	4.407	0.4~0.5 [7]
KCl:I$^-$ (0.2 mol.%)	174.7~175.8	0.037~0.046	4.65~5.19	1.01	4.407	0.4~0.5 [7]
RbCl:Br$^-$ (1.0 mol.%)	174.0	0.053	4.36	0.776	4.609	0.50 [8]
RbCl:I$^-$ (1.0 mol.%)	173.0	0.061	4.05	0.776	4.609	0.58 [8]

*The values of shear modulus in Table 1 are approximately estimated as each matrix of various crystals at about 0 K.

As shown in Table 1, these calculations for all crystals are within $173° \leq \phi_0 < 180°$, where the Friedel relation is applicable to the interaction between a dislocation and the monovalent anions in the halide crystals. Then, the values of $(L_0/L)^2$ are derived from combining equations (1) and (2), namely

$$\left(\frac{L_0}{L}\right)^2 = \cos\frac{\phi_0}{2} \tag{7}$$

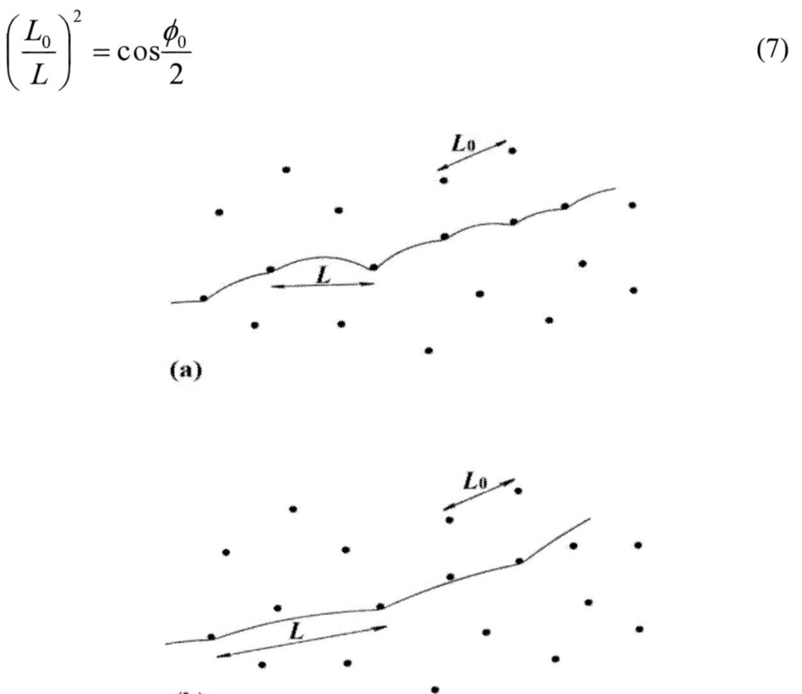

Figure 2. Relation between L and L_0 for the case: the breaking angle is (a) small (strong as obstacles to the dislocation motion) and (b) large (weak as obstacles to it). The solid circles (.) represent localized obstacles.

It can be found within $0.029 \leq (L_0/L)^2 \leq 0.061$, in other words, $4.05 \leq (L/L_0) \leq 5.87$ in Table 1. The value of (L/L_0) becomes lower at lower ϕ_0 as given in Table 1. Small bending angle ϕ_0 suggests that the additive anion is strong as an obstacle to dislocation motion (I^- ions in RbCl are the strongest of all crystals here). This leads to the result that L approaches to the length of L_0 by the curvature of dislocation as shown in figure 2: (a) strong and (b) weak as obstacles to the dislocation motion.

CONCLUSION

The breaking angle ϕ_0 of a dislocation ranges 173° to 177° by Cl⁻, Br⁻ or I⁻ in the various alkali halides single crystals (NaCl:Br⁻, NaBr: Cl⁻ or I⁻, KCl: Br⁻ or I⁻, and RbCl: Br⁻ or I⁻). This suggests that the additive anions are the weak obstacles such as impede the mobile dislocation at $\phi_0 \approx 180$ degrees. Then, the values of (L/L_0) were found to be within 4.05 to 5.87 there.

REFERENCES

[1]   Friedel, J. *Dislocations*; Pergamon Press; Oxford, 1964; p 224.

[2]   Kohzuki, Y. *J. Mater. Sci.* 2000, *35*, 3397-3401.

[3]   Igata, N. *Strength of Materials*; Baifukan; Tokyo, 1991; pp 8-12 (in Japanese).

[4]   Sirdeshmukh, D. B.; Sirdeshmukh, L.; Subhadra, K. G. *Alkali Halides*, (edited by Hull, R.; Osgood, Jr. R. M.; Sakaki, H.; Zunger, A.) Springer-Verlag; Berlin Heidelberg, 2001; p 7.

[5]   Ohgaku, T.; Teraji, H. *Phys. Status Solidi* (a) 2001, *187*, 407-413.

[6]   Ohgaku, T.; Sakuda, M. *Mater. Sci. Engng. A* 2003, *357*, 55-60.

[7]   Ohgaku, T.; Takeuchi, N. *Meeting abstracts of the Physical Society of Japan* Vol. 1991, No. 2 (199110311) p 54 (in Japanese).

[8]   Ohgaku, T.; Hashimoto, K. *Mater. Sci. Engng. A* 2005, *400-401*, 401-404.

In: Halides
Editor: Jean Lefebure

ISBN: 978-1-62417-947-1
© 2013 Nova Science Publishers, Inc.

Chapter 5

THE NONLINEAR ELASTIC PROPERTIES AND GRÜNEISEN CONSTANTS OF ALKALI HALIDE[*]

X. Z. Wu[†], L. L. Liu, R. Wang, C. B. Li, S. H. Wu and H. F. Feng
College of Physics and Institute for Structure and Function,
Chongqing University, Chongqing, P. R. China

Abstract

The second-order elastic constants (SOECs) and third-order elastic constants (TOECs) of the alkali halides (LiF, NaF, KF, LiCl, NaCl and KCl) with NaCl structure is presented using the density functional theory (DFT) and homogeneous deformation method. The k-points and the cutoff energy are given reliable verification while TOECs are concerned. From the nonlinear least-square fitting, the elastic constants are extracted from a polynomial fit to the calculated strain-energy data. Both the SOECs and TOECs are in agreement with the available experimental and previous results. The Lagrangian stress is changing linearly for small Lagrangian strain and it is found that the nonlinear effects play an important role while the finite strains are larger than approximately 0.025. The pressure derivatives of SOECs are calculated directly using the TOECs, which is used to discuss roughly the phase transition pressure of alkali halide. The Grüneisen constants of long-wavelength acoustic modes characterizing the anharmonic properties of alkali halide are also presented.

---

[*]Project Supported by the Natural Science Foundation of China (11104361).
[†]E-mail address: xiaozhiwu@cqu.edu.cn

Keywords: Elastic properties, alkali halide, Grüneisen constants, first principle study

1. Introduction

In the theory of linear elasticity, the infinitesimal deformations are assumed. As a result, the second-order elastic constants (SOECs) are sufficient to describe the linear elastic stress-strain response and the propagation velocity of acoustic disturbances along different crystallographic directions [1, 2]. In the case of finite strains, the theory of nonlinear elasticity is required [3, 4, 5, 6, 7]. The third-order elastic constants (TOECs) play an important role as well as SOECs [6]. TOECs are useful not only in describing mechanical phenomena when applying larger stress and strain, but they can also serve as a basis for describing the anharmonic properties such as thermal expansion, temperature dependence of elastic properties, phonon-phonon interaction, Grüneisen parameter etc [8, 9].

Many experiments have been performed to determine SOECs and TOECs. However, it is still rather difficult to obtain a complete set of TOECs through the experiment [9]. As a result, several theoretical approaches have been introduced to calculate the TOECs. These methods include empirical interatomic force-constant model [10, 11], molecular-dynamics simulations using fluctuation formulas [12, 13], and the first-principles quantum mechanics calculations [14, 15]. The use of first-principles quantum mechanics calculations method to determine TOECs was first introduced by Nielson and Martin [14, 15]. Employing the stress theorem [16] within density functional theory, they employed the method of homogeneous deformation combined with first-principles quantum mechanics calculation to determine the SOECs and TOECs for several crystals having the diamond structure. Recently, the methods of first-principles calculations have been employed to obtain TOECs through applying series of finite deformations [17, 18, 19, 20, 21]. Their results are agree well with experiments.

In this work, we shall employ the method of homogeneous deformation combined with first-principle internal energy-strain calculations to provide a mode of determining TOECs for alkali halide crystals LiF, NaF, KF, LiCl, NaCl, KCl with NaCl (B1)-type structure. Besides, the homogeneous deformation strain used are usually simple deformation modes such as uniaxial tensor or compression, simple or pure shear, and other combinations of homogeneous stains. Alkali halides are substances of interest from the theoretical point of view because a simple model of their structure has been quite successful. The alkali halides are the simplest ionic crystals and measurement of their elastic properties have been very useful for short range interatomic forces and anharmonic properties. The TOECs for alkali halide crystals have been a topic of extensive study for the past few decades [3, 7, 22, 23, 24, 25, 26, 27, 28, 29] and there exist many experimental and theoretical works on the TOECs for the ionic crystals using the atomistic model. Although atomistic calculations of the elastic constants have become fairly commonplace in the last few decades and the methods of calculation almost routine, there are many aspects of these calculations which are still either controversial or require extensive development. In this paper, we perform ab initio calculations of the alkali halide crystals LiF, NaF, KF, LiCl, NaCl and KCl with B1 structure in order to better understand the strain behavior in these ionic crystals. Besides, we also briefly discuss the applications of c_{ijk} to determine other anharmonic properties, namely, pressure derivatives of second-order elastic constants c'_{ij} and mode Grüneisen constants $\gamma(\mathbf{q}, \mathbf{j})$.

The paper is organized as follows. In section 2, we give a general overview of the nonlinear elasticity theory. Section 3 contains a description of methodology and results discussion. Section 4 and 5 deal with the determination of quantities third-order elastic constants, namely, the pressure derivatives of second-order elastic constant and mode Grüneisen constants, respectively. Section 6 is a conclusion.

2. Theory of Nonlinear Elasticity

Here, we will recall some basic facts from nonlinear theory of elasticity [4, 6, 7]. Let us consider the initial point with the coordinate a_i which, after applying a finite deformation to a crystal, moves to the position x_i. The deformation applied to the crystal is described by the deformation gradient

$$F_{ij} = \frac{\partial x_i}{\partial a_j}. \tag{1}$$

where i and $j(=1,2,3)$ represent the Cartesian coordinates. Then the Lagrangian strain η [6] is defined as

$$\eta_{ij} = \frac{1}{2}\sum_p (F_{pi}F_{pj} - \delta_{ij}). \tag{2}$$

Because the Lagrangian strains are symmetric, they contain no information regarding rigid rotation of the crystal element.

The elastic constants are defined by expanding the internal energy U as a Taylor series in elastic strain at constant entropy by Brugger [7]

$$U(V,\eta,S) = U(V,0,S) + \frac{V}{2!}\sum_{ijkl} C^S_{ijkl}\eta_{ij}\eta_{kl} + \frac{V}{3!}\sum_{ijklmn} C^S_{ijklmn}\eta_{ij}\eta_{kl}\eta_{mn} + \dots, \quad (3)$$

where S is the entropy, V is the volume of the unstrained crystal, and $U(V,0,S)$ is the corresponding internal energy of ground state. An expansion in terms of symmetric strains is appropriate due to the internal energy is invariant under rigid rotation [30]. The second, third and higher orders elastic constants are defined as the second-, third-, and higher- orders derivatives of the internal energy with respect to the strain, respectively. So, the isentropic elastic constants are

$$C^S_{ijkl} = \frac{1}{V}\frac{\partial^2 U}{\partial\eta_{ij}\partial\eta_{kl}}\bigg|_{\eta=0} \qquad \text{(SOEC).} \qquad (4)$$

$$C^S_{ijklmn} = \frac{1}{V}\frac{\partial^3 U}{\partial\eta_{ij}\partial\eta_{kl}\partial\eta_{mn}}\bigg|_{\eta=0} \qquad \text{(TOEC).} \qquad (5)$$

Similarly, isothermal elastic constants are obtained by expanding the Helmholtz free energy as a Taylor series in Lagrangian strain tensor η at constant temperature [30, 31]

$$F(V,\eta,T) = U(V,0,T) + \frac{V}{2!}\sum_{ijkl} C^T_{ijkl}\eta_{ij}\eta_{kl} + \frac{V}{3!}\sum_{ijklmn} C^T_{ijklmn}\eta_{ij}\eta_{kl}\eta_{mn} + \dots, \quad (6)$$

In our first-principles calculations, deformations are applied under isothermal conditions $T = 0K$, $F = U - TS = U$, so $C^S = C^T$. We will not distinguish those two types of elastic constants in the following.

Since the Lagrangian strain tensor is symmetric, i.e., $\eta_{ij}=\eta_{ji}$, only six of each set of nine variables are independent, and we can simplify the notations in the tensors by using the Voigt notation (xx→1, yy→2, zz→3, yz→4, xz→5, xy→6). We have therefore for the Lagrangian strain tensors η as follows

$$\eta = \begin{pmatrix} \eta_1 & \frac{\eta_6}{2} & \frac{\eta_5}{2} \\ \frac{\eta_6}{2} & \eta_2 & \frac{\eta_4}{2} \\ \frac{\eta_5}{2} & \frac{\eta_4}{2} & \eta_3 \end{pmatrix} \qquad (7)$$

Equations (3) and (6) now can be written as

$$V^{-1}[U(V,\eta) - U(V,0)] = \frac{1}{2!} \sum_{i,j=1,6} c_{ij}\eta_i\eta_j + \frac{1}{3!} \sum_{i,j,k=1,6} c_{ijk}\eta_i\eta_j\eta_k + \ldots, \qquad (8)$$

For our researched alkali halide crystals with B1 structure are cubic symmetry, we can express Eq.(8) in the second-, third-order terms of the strain tensor [17, 19]

$$V^{-1}[U(V,\eta) - U(V,0)] = \psi_2 + \psi_3 + \ldots, \qquad (9)$$

where

$$\psi_2 = \tfrac{1}{2}c_{11}(\eta_1^2 + \eta_2^2 + \eta_3^2) + c_{12}(\eta_1\eta_2 + \eta_2\eta_3 + \eta_3\eta_1)$$
$$+ \tfrac{1}{2}c_{44}(\eta_4^2 + \eta_5^2 + \eta_6^2), \qquad (10)$$

$$\psi_3 = \tfrac{1}{6}c_{111}(\eta_1^3 + \eta_2^3 + \eta_3^3)$$
$$+ \tfrac{1}{2}c_{112}(\eta_2\eta_1^2 + \eta_3\eta_1^2 + \eta_1\eta_2^2 + \eta_1\eta_3^2 + \eta_2\eta_3^2 + \eta_3\eta_2^2)$$
$$+ c_{123}\eta_1\eta_2\eta_3 + \tfrac{1}{2}c_{144}(\eta_1\eta_4^2 + \eta_2\eta_5^2 + \eta_3\eta_6^2) \qquad (11)$$
$$+ \tfrac{1}{2}c_{155}(\eta_2\eta_4^2 + \eta_3\eta_4^2 + \eta_1\eta_5^2 + \eta_3\eta_5^2 + \eta_1\eta_6^2 + \eta_2\eta_6^2)$$
$$+ c_{456}\eta_4\eta_5\eta_6$$

As shown above, there are three independent SOECs (c_{11}, c_{12}, c_{44}) and six TOECs (c_{111}, c_{112}, c_{123}, c_{144}, c_{155}, and c_{456}). As shown below, these elastic constants can be calculated by subjecting the system to various simple Lagrangian strain tensors η, provided also that the system has no any polymorphic phase transition and internal deformation under these strains.

Another fundamental quantity in the theory of nonlinear elasticity is Lagrangian stress, which is defined as the first-order derivative of the internal energy or Helmholtz free energy with respect to the Lagrangian strain tensor,

$$\sigma_{ij} = V^{-1}(\partial U/\partial \eta_{ij}) = V^{-1}(\partial F/\partial \eta_{ij}) \qquad (12)$$

which can also be used to calculate the elastic constants [14, 15, 17, 19].

Table 1. The coefficients P_2 and P_3 in Eq. (13) of corresponding Lagrangian strains are shown as combinations of second- and third-order elastic constants for cubic crystals

strain type	P_2	P_3
$\eta_A=(\xi,0,0,0,0,0)$	$\frac{1}{2}c_{11}$	$\frac{1}{6}c_{111}$
$\eta_B=(\xi,\xi,0,0,0,0)$	$c_{11}+c_{12}$	$\frac{1}{3}c_{111}+c_{112}$
$\eta_C=(\xi,\xi,\xi,0,0,0)$	$\frac{3}{2}c_{11}+3c_{12}$	$\frac{1}{2}c_{111}+3c_{112}+c_{123}$
$\eta_D=(\xi,0,0,\xi,0,0)$	$\frac{1}{2}c_{11}+\frac{1}{2}c_{44}$	$\frac{1}{6}c_{111}+\frac{1}{2}c_{144}$
$\eta_E=(\xi,0,0,0,0,2\xi)$	$\frac{1}{2}c_{11}+2c_{44}$	$\frac{1}{6}c_{111}+2c_{166}$
$\eta_F=(0,0,0,\xi,\xi,\xi)$	$\frac{3}{2}c_{44}$	c_{456}

3.   Determination of the Elastic Constants

3.1.   Methodology and Computational Details

In this work, we have determined second- and third-order elastic constants for alkali halide crystals LiF, NaF, KF, LiCl, NaCl, KCl with B1 structure on the basis of quantum density functional theory (DFT) calculations for homogeneous deformation crystals. Eqs. (8), (10) and (11) shows that we can treat the internal energy variations as a polynomial function of the Lagrangian strain η_{ij}. Since the Lagrangian strain tensor is symmetric, i.e., $\eta_{ij}=\eta_{ji}$, only six of each set of nine variables are independent. We can select certain simple deformation or loading modes such as the Lagrangian tensor η only has one or a few components and the nonzero strain components of each Lagrangian strain tensor in terms of a single parameter ξ. To obtain a solvable system for the TOECs, the number of applied Lagrangian strain tensors must be as large as the number of independent TOECs of the crystal. Our researched alkali halide crystals with B1 structure are cubic symmetry, we need consider six sets of deformations. The selection of six different deformation strain leads to different strains used in this work, which are labeled as η_α, $\alpha=A$, B, ..., F and listed in Table 1. Inserting these strains into Eqs. (8)-(12), the internal energy per unit mass can be written as an expansion in the strain parameter ξ [19].

$$V^{-1}[U(V, \eta) - U(V, 0)] = f_\alpha[\eta_\alpha(\xi)]$$

$$= P_2\xi^2 + P_3\xi^3 + O(\xi^4), \tag{13}$$

where the coefficients P_2 and P_3 are combinations of second- and third-order elastic constants of the crystal, respectively. The relationship between the coefficients P_2, P_3 and the second- and third-order elastic constants are presented in Table 1 for the specific strain tensors listed also in the Table 1.

To implement the different deformation modes in our calculation, we need to have the deformation gradient matrix $\mathbf{F}$, which can be obtained from the Lagrangian strain η by inverting Eq. (2) [19],

$$F_{ij} = \delta_{ij} + \eta_{ij} - \frac{1}{2}\sum_k \eta_{ki}\eta_{kj} + \ldots, \tag{14}$$

For a given η, in general, the associated deformation gradient matrix $\mathbf{F}$ is not unique, while the various possible solutions differing from one another by a rigid rotation. The lack of a one-to-one relationship between the Lagrangian strain η and deformation gradient matrix $\mathbf{F}$ is not a problem because the Lagrange strain brings rotational invariance of total energy [18, 19]. Furthermore, Eq. (14) provides a unique relation between η and $\mathbf{F}$ for the system without rigid rotation. To obtain the unit cell of strained crystal, the deformation gradient matrix $\mathbf{F}$ is applied to the undeformed primitive vector $\mathbf{a}_i$, the deformed or strained primitive vector $\mathbf{a}'_i$,

$$\mathbf{a}'_i = \sum_j F_{ij}\mathbf{a}_j \tag{15}$$

For each strain tensor chosen in our calculation labeled as $\alpha=A$, B, $\ldots$, F listed in Table 1, we change the value of ξ from $-|\xi_{max}|$ to $+|\xi_{max}|$ with a finite step size 0.008, the maximum strain parameter ξ_{max} included in the polynomial fit is an important parameter in these calculations. In our work, the fitted coefficient P_2 for the combination of the second-order elastic constants is very stable and is almost independent of the range of fitting. However, the coefficient P_3 is more sensitive to ξ_{max}. To illustrate this feature, several TOECs of KCl are plotted in Fig. 1 as a function of the ξ_{max}. As shown, the results of the TOECs from the polynomial fit are well converged after the maximum strain parameters ξ_{max} reaches 0.06. Therefore, the $\xi_{max}=0.08$ employed in this work for our researched alkali halide crystals is sufficient.

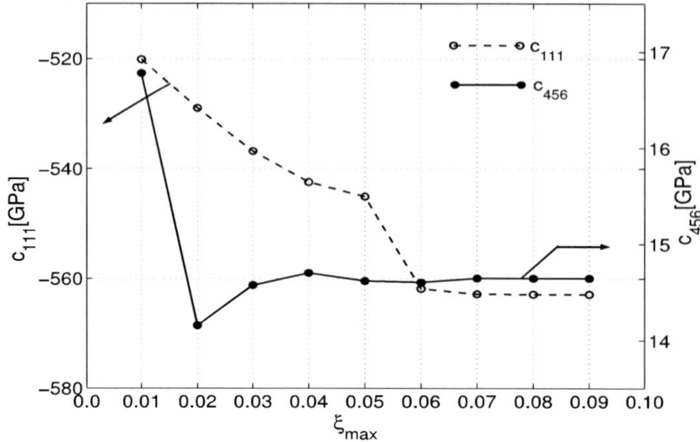

Figure 1. Sample convergence tests for the third-order elastic constants in KCl as a function of the maximum strain $|\xi_{max}|$. Note different scales for c_{111} and c_{456}.

We carry out first-principle total-energy calculations in the framework of the DFT, using the Vienna ab initio simulation package (VASP 4.6) developed at the Institut für Materialphysik of Universität Wien [32, 33, 34]. For the exchange-correlation functional, generalized gradient approximation (GGA) according to Perdew-Burke-Ernzerhof (PBE) [35, 36] has been applied. The projector augmented wave (PAW) method [37] has been employed in its variant available in the VASP package [38]. Since high accuracy is needed to evaluate the TOECs, to test the convergence of the total energy and the TOECs, we employ the k-point mesh up to $19 \times 19 \times 19$ in our calculations following the Monkhost-Pack scheme [39]. The structures are relaxed without any symmetry constraints with a cutoff energy of 500 eV for our researched alkali halides. The k-points grid size and the cutoff energy employed here are very reliable for LiF, NaF, KF, LiCl, NaCl and KCl on the basis of our tests. A sufficiently large Fourier grid including all wave-vectors up to twice the cutoff wavenumber is used to avoid any wraparound errors. The equilibrium theoretical lattice structures are obtained by minimizing the Hellmann-Feynman force on the atoms and the stress on the unit cell. The convergence of energy and force is set as 1.0×10^{-6} eV and 1.0×10^{-4} eV/Å, respectively. In Table 2, we give the all equilibrium lattice constants for all studied alkali halide crystals in our calculation, and it is shown that

Table 2. Calculated equilibrium lattice constants for the alkali halides in our calculation comparison with the available experiment data. The unit of all data is Å

	LiF	NaF	KF	LiCl	NaCl	KCl
This work	4.06	4.67	5.42	5.15	5.68	6.38
Experiment [40, 41]	4.03	4.63	5.34	5.14	5.64	6.30

the results agree well with the experiment values. In addition, we would like to mention that since our calculation is performed at 0 K and the experimental results are measured at room temperature, certain differences in the lattice constants should be anticipated. Taking NaCl as an example, we show the internal energy calculations for bulk NaCl in relation to the lattice constants a in Fig. 2.

Besides, the third-order elastic constants (TOECs) are also very sensitive to the parameters of the k-point gird size and the cutoff energy. Fig. 3 show the convergence test for the elastic constants of KCl. It illustrates the dependence of two sample third-order elastic constants c_{111} and c_{456} on the k-point grid size and cutoff energy of KCl. For the employed parameters ($E_{\text{cutoff}}^{\text{KCl}} = 500$ eV and $19\times19\times19$ k-point grid size) in our calculations, the relative difference between two successive values of examined constants in our test is lower than 1.0%. The accuracy for the TOECs is very reasonable for elastic applications. This similar tests have also been applied for other alkali halides, and the same cutoff energy 500 eV and k-point mesh size $19 \times 19 \times 19$ are sufficiently large for the internal energy to converge to the equilibrium state and for the TOECs to converge well.

3.2.   Results and Discussion

Table 3-8 give our prediction for the values of TOECs for LiF, NaF, KF, LiCl, NaCl and KCl, respectively. For completeness, we also provide there our prediction for second-order elastic constants and compare them with previous calculations and experiment. For c_{ij} values, sometimes it is possible to determine one constant from a few fits [e.g., c_{44} from coefficients in $f_D(\xi)$, $f_E(\xi)$, and $f_F(\xi)$], obtaining slightly different results [e.g., for KCl, c_{44}=64.6, and 65.3, and 65.1 GPa from $f_D(\xi)$, $f_E(\xi)$, and $f_F(\xi)$, respectively]. In such cases, the average of all

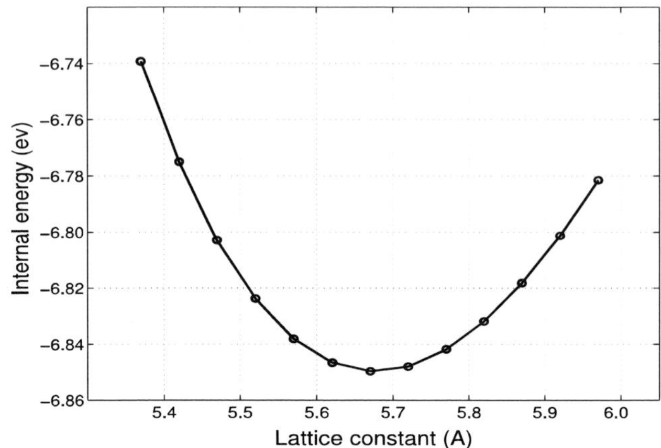

Figure 2. The calculated internal energy of NaCl as a function of lattice constant, the equilibrium lattice parameter is found to be 5.68 Å, which is determined from the corresponding minimum value of the internal energy.

obtained values are given in Table 3-8. It is easily found that the present values of SOECs are in good agreement with the available data from either recent theoretical calculation or experiment values. The E-ξ curves for KCl included the results of the first-principles calculations and the fitted polynomial are shown in Fig. 4. The relations between E and and ξ are fitted with a fourth-order polynomial and the fitted coefficients in the polynomials are obtained numerically. Employing relations in Table 1, we can easily obtain the elastic constants listed in Table 3-8.

As far as calculations of third-order elastic constants are concerned, they also require extremely good convergence of parameters included cutoff energy and k-points gird size governing the accuracy of computations. The convergence tests of k-point grid size and cutoff energy (see in Fig. 3) show that we give very reliable information of the TOECs for our researched alkali halide crystals with B1 structure. The usage of PAW formalism chosen to solve the Kohn-Sham (KS) equations seems not to influence the results significantly [17], because the calculations of the static and dynamical properties for a broad range solids within PAW has been performed properly [54]. In the present calculations, we use the GGA-PBE exchange-correlation functional which is commonly considered to be one of the best in the market [17]. For comparison, we easily

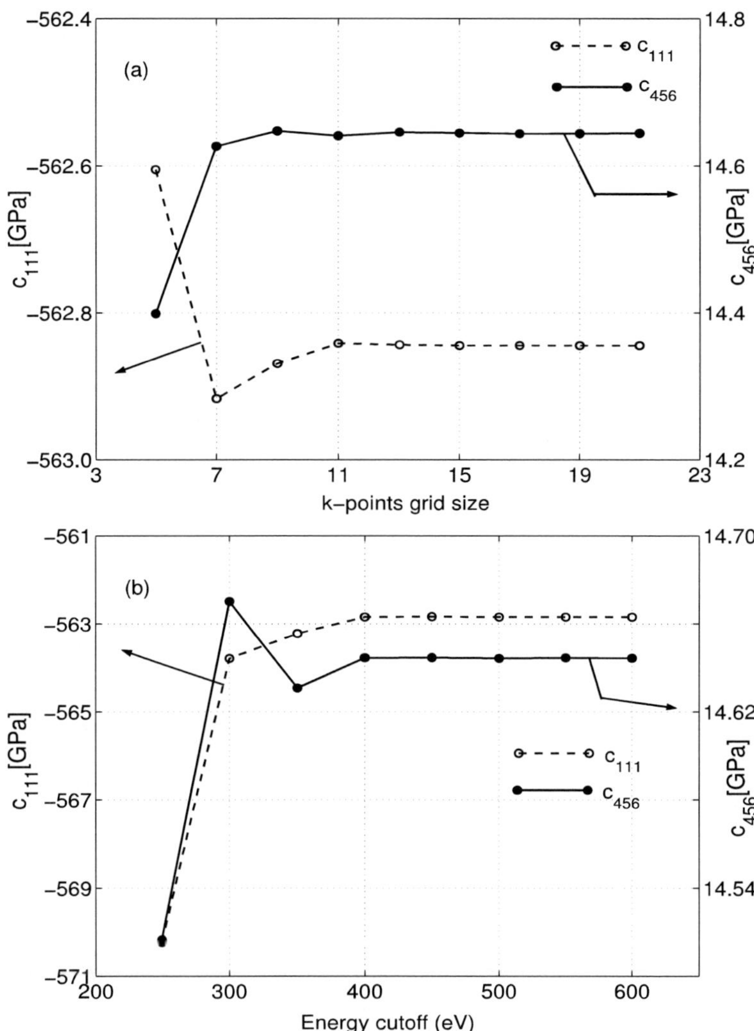

Figure 3. Sample convergence tests for the TOECs in KCl. (a) The dependence of the TOECs c_{111} and c_{456} on the density of k-points mesh (energy cutoff of 500 eV is applied for all points). (b) The dependence of the TOECs c_{111} and c_{456} on the cutoff energy (Monkhorst-Pack sampling $19 \times 19 \times 19$ is used for all points).

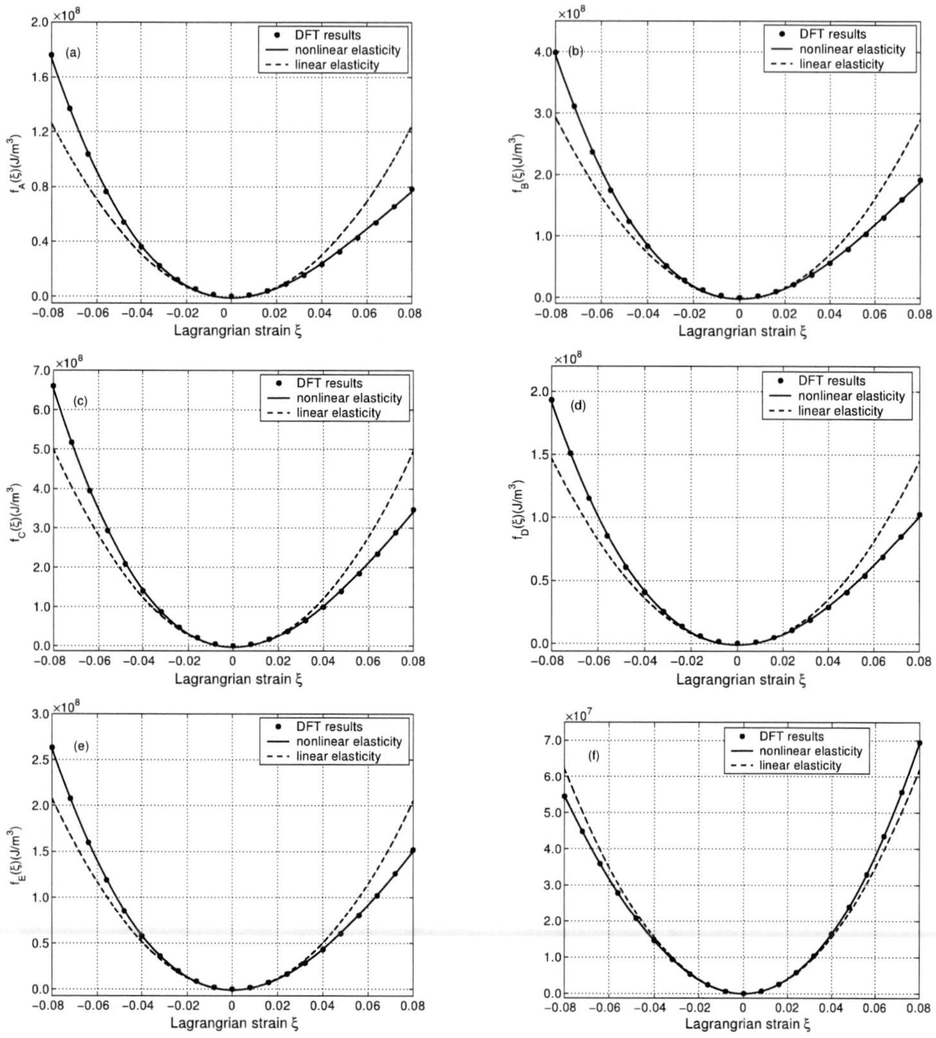

Figure 4. The strain energy relations as a function of linear strain parameter ξ for KCl cubic crystal. Circle points denote results of DFT computations; solid and dashed lines represent the curves obtained from nonlinear and linear elasticity theory, respectively. (a), (b), (c), (d), (e), and (f) describe Lagrangian strains η_A, η_B, η_C, η_D, η_E, and η_F, respectively.

Table 3. Calculated elastic constants for LiF in our calculation comparison with the previous study and available experiment data. The unit of all data is 10^{11}N/m^2

	Present results	Previous study	Experiment
LiF			
c_{11}	1.118	1.139[a], 1.112[c], 1.137[d], 1.195[e]	1.105[b]
c_{12}	0.510	0.477[a], 0.420[c], 0.476[d], 0.504[e]	0.443[b]
c_{44}	0.616	0.636[a], 0.628[c], 0.637[d], 0.504[e]	0.637[b]
c_{111}	-14.244	-13.530[b], -22.300[f], -14.149[g], -15.20[h]	-14.230±0.300[a]
c_{112}	-1.244	-2.495[b], -1.900[f], -2.481[g], -3.300[h]	-2.640±0.230[a]
c_{123}	0.868	0.879[b], 0.870[f], 0.801[g], 1.900[h]	1.560±0.280[a]
c_{144}	0.846	0.980[b], 1.000[f], 0.867[g], 0.900[h]	0.850±0.100[a]
c_{166}	-2.350	-2.542[b], -1.780[f], -2.411[g], -3.130[h]	-2.730±0.130[a]
c_{456}	0.899	0.955[b], 1.000[f], 0.932[g], 0.860[h]	0.940±0.060[a]

[a] see Ref. [42],   [b] see Ref. [23]
[c] see Ref. [43],   [d] see Ref. [44]
[e] see Ref. [45],   [f] see Ref. [46]
[g] see Ref. [47],   [h] see Ref. [48]

find that our results of the TOECs for LiF, NaF, KF, LiCl, NaCl and KCl have good agreement with the recent theoretical and available experimental values. In a word, we give very reliable information of the elastic constants for our researched alkali halide crystals.

Next interesting issue is to examine for which range of deformations the third-order effects dominant the properties of solids. In Fig. 4, we compare energy values obtained within linear and nonlinear elasticity and DFT results for six sets of deformation in KCl crystal. Besides we can calculate the Lagrangian stress σ when we substitute the obtained elastic constants into Eq. (12). In Fig. 5, we also show the Lagrangian stress curves of the linear elasticity comparison with the nonlinear theory for the particular deformation η_C and η_F in KCl, respectively. It is apparent to see that the linear elasticity is not sufficient for strains lager than approximately 2.5% and the third-order effects must be considered. It is also worth noting that for all studied alkali halide crystals and examined Lagrangian strains up to 8.0%, including the terms up to third-order in energy expansion [Eq. (8)] sufficed to obtain well agreement with DFT results.

Table 4. Calculated elastic consants for NaF in our calculation comparison with the previous study and available experiment data. The unit of all data is 10^{11}N/m^2

	Present results	Previous study	Experiment
NaF			
c_{11}	0.967	0.963[a], 0.970[b], 0.970[c], 0.850[d]	0.945[e]
c_{12}	0.246	0.240[a], 0.248[b], 0.236[c], 0.288[d]	0.213[e]
c_{44}	0.306	0.276[a], 0.282[b], 0.282[c], 0.288[d]	0.282[e]
c_{111}	-1.296	-8.563[e], -7.980[f], -1.983[g], -1.488[h], -1.510[s]	-1.480±0.800[a]
c_{112}	-0.473	-1.394[e], -1.060[f], -0.650[g], -1.927[h], -2.200[s]	-2.700±0.400[a]
c_{123}	0.393	0.488[e], 1.020[f], 0.230[g], 0.443[h], 3.500[s]	2.860±0.600[a]
c_{144}	0.470	0.561[e], 0.570[f], 0.560[g], 0.467[h], 0.500[s]	0.460±0.060[a]
c_{166}	-1.133	-1.429[e], -1.000[f], -1.360[g], -1.115[h], -1.230[s]	-1.140±0.030[a]
c_{456}	0.557	0.544[e], 0.570[f], 0.590[g], 0.460[h], 0.500[s]	0.000±0.010[a]

[a]see Ref. [49],   [b]see Ref. [44],   [c]see Ref. [50]
[d]see Ref. [45],   [e]see Ref. [23],   [f]see Ref. [46]
[g]see Ref. [28],   [h]see Ref. [47],   [s]see Ref. [48]

4.   The Effective Elastic Constants under Pressure

The alkali halide crystals are the simplest ionic solids. In the case of materials under hydrostatic large pressure, it is of value to describe the nonlinear elastic properties employing the concept of effective elastic constants $c_{ij}(P)$. For many applications, considering only linear terms in the external hydrostatic pressure is sufficient [3, 17]

$$c_{ij}(P) \approx c_{ij} + c'_{ij}P, \tag{16}$$

where pressure derivative c'_{ij} is a material-dependent parameter. Naturally, the values of c'_{ij} can be recovered from both second-order elastic constants (SOECs) and third-order elastic constants (THECs). The necessary formulas are given

Table 5. Calculated elastic constants for KF in our calculation comparison with the previous study and available experiment data. The unit of all data is 10^{11}N/m^2

	Present results	Previous study	Experiment
KF			
c_{11}	0.584	0.195[a], 0.581[b]	0.656[c]
c_{12}	0.136	0.171[a], 0.163[b]	0.146[c]
c_{44}	0.135	0.193[a], 0.163[b]	0.125[c]
c_{111}	-9.259	-5.715[a], -8.800[d]	-7.622[c]
c_{112}	-0.379	-0.842[a], -0.990[d]	-0.713[c]
c_{123}	0.255	0.228[a], 0.300[d]	0.242[c]
c_{144}	0.299	0.340[a], 0.310[d]	0.316[c]
c_{166}	-0.749	-0.890[a], -0.640[d]	-0.753[c]
c_{456}	0.309	0.315[a], 0.410[d]	0.306[c]

[a] see Ref. [27],   [b] see Ref. [45]
[c] see Ref. [23],   [d] see Ref. [48]

below [3]

$$c'_{11} = -\frac{2c_{112} + c_{111} + 2c_{12} + 2c_{11}}{2c_{12} + c_{11}},$$

$$c'_{12} = -\frac{2c_{112} + c_{123} - c_{12} - c_{11}}{2c_{12} + c_{11}},$$

$$c'_{44} = -\frac{2c_{166} + c_{144} + c_{44} + 2c_{12} + c_{11}}{2c_{12} + c_{11}}. \tag{17}$$

Results for the pressure derivatives c'_{ij} calculated on the basis of our prediction for second- and third-order elastic constants are shown in Table 9. In Ref. [59], the following method has been employed to determine the pressure dependence of the second-order elastic constants. First, applying the hydrostatic to a crystal, and then the crystal has been additionally deformed to obtain the pressure dependent elastic constants. The ab initio calculated results for the total elastic energy combined with the strain-energy relation will enable us to determine $c_{ij}(P)$ as well as c'_{ij}. Furthermore, the methods of calculations on the pressure derivatives have also been used in Ref. [17] for studying the related properties

Table 6. Calculated elastic constants constants for LiCl in our calculation comparison with the previous study and available experiment data. The unit of all data is 10^{11}N/m^2

	Present results	Previous study	Experiment
LiCl			
c_{11}	0.532	0.421[a]	0.494[b]
c_{12}	0.223	0.192[a]	0.228[b]
c_{44}	0.247	0.192[a]	0.246[b]
c_{111}	-7.101	-9.500[c]	-8.572[b]
c_{112}	-0.673	-1.000[c]	-0.859[b]
c_{123}	0.567	0.500[c]	0.294[b]
c_{144}	0.338	0.370[c]	0.374[b]
c_{166}	-0.940	-1.510[c]	-0.901[b]
c_{456}	0.268	0.300[c]	0.362[b]

[a]see Ref. [45]

[b]see Ref. [23]

[c]see Ref. [48]

of selected semiconductors. Therefore, we believe that the approach employed in the present paper is accurate.

As is well known, the mechanical stability of a cubic crystal under isotropic pressure is judged by the following relation [71]

$$c_{11} + 2c_{12} > 0, \quad c_{11} - c_{12} > 0, \quad c_{44} > 0. \tag{18}$$

The B1 phase is mechanically unstable and the structural transformation occurs at the phase transition pressure. Unfortunately, the mechanical criterion is failed for alkali halide except KCl. However, the phase transition pressure of KCl obtained by mechanical criterion is 18.2GPa, which is far higher than the pressure of about 2.0 GPa obtained by equating the enthalpies of the B1 and B2 phase or the experimental method (see Table 10). This large discrepancy predicted by the elastic instability criterion and the thermodynamic criteria is also existed for CaO [71], which can be interpreted as the presence of a substantial activation barrier to this particular transition route [72].

Table 7. Calculated elastic constants for NaCl in our calculation comparison with the previous study and available experiment data. The unit of all data is 10^{11}N/m^2

	Present results	Previous study	Experiment
NaCl			
c_{11}	0.477	0.493[a], 0.498[b], 0.494±0.0006[c], 0.487[d], 0.358[e], 0.479[f],	0.479[g]
c_{12}	0.122	0.129[a], 0.130[b], 0.127±0.0003[c], 0.124[d], 0.098[e], 0.126[f]	0.115[g]
c_{44}	0.125	0.128[a], 0.128[b], 0.128±0.0001[c], 0.126[d], 0.123[e], 0.126[f]	0.127[g]
c_{111}	-6.884	-8.800[a], -8.230±0.200[b], -8.605[g], -7.364[e], -8.100[h]	-8.430±0.330[c]
c_{112}	-0.287	-0.571[a], -0.500±0.020[b], -0.522[g], -0.530[e], -0.660[h]	-0.500±0.070[c]
c_{123}	0.259	0.284[a], 0.530±0.070[b], 0.164[g], 0.130[e], 0.450[h]	0.460±0.090[c]
c_{144}	0.215	0.257±0.160[a], 0.230±0.030[b], 0.256[g], 0.273[e], 0.320[h]	0.290±0.050[c]
c_{166}	-0.618	-0.611±0.007[a], -0.610±0.030[b], -0.574[g], -0.605[e], -0.710[h]	-0.600±0.040[c]
c_{456}	0.213	0.271±0.014[a], 0.200±0.010[b], 0.248[g], 0.253[e], 0.350[h]	0.260±0.010[c]

[a]see Ref. [25],   [b]see Ref. [51]

[c]see Ref. [42],   [d]see Ref. [52]

[e]see Ref. [27],   [f]see Ref. [45]

[g]see Ref. [23],   [h]see Ref. [48]

5.   Anderson-Grüneisen Parameter

It is a more normal practice to characterize anharmonic properties of crystals in terms of generalized Grüneisen parameters that describe the strain dependence of the mode frequencies and are defined by [9, 73]

$$\gamma(\mathbf{q}, \mathbf{j}) = -\frac{V}{\omega_{\mathbf{q}\mathbf{j}}} \frac{\partial}{\partial V} \omega_{\mathbf{q}\mathbf{j}(V)}, \tag{19}$$

where V is the crystal volume, $\mathbf{q}$ denotes the phonon-wave vector and $\mathbf{j}$ is the polarization vector.

Based on the continuum limit, one may express mode Grüneisen constants of long-wavelength acoustic modes in terms of second- and third-order elastic moduli. It is found that the Grüneisen constants depend strongly on those elastic constants. Using the Grüneisen parameter expressions [9, 73] and the elastic constants calculated employing the method of homogeneous deformation combined with first-principle strain-energy estimates, we have calculated the mode Grüneisen constants and listed the results of $\gamma(\mathbf{q}, \mathbf{j})$ in Table 11. To the best of our knowledge, there are no available experimental data of $\gamma(\mathbf{q}, \mathbf{j})$ even to date,

Table 8. Calculated elastic constants for KCl in our calculation comparison with the previous study and available experiment data. The unit of all data is 10^{11}N/m^2

	Present results	Previous study	Experiment
KCl			
c_{11}	0.361	0.408^a, 0.409^b, 0.403^c, 0.375^d	0.398^e
c_{12}	0.061	0.071^a, 0.070^b, 0.066^c, 0.084^d	0.061^e
c_{44}	0.065	0.063^a, 0.062^b, 0.064^c, 0.084^d	0.063^e
c_{111}	-5.628	-7.010^a, -6.013^e, -5.461^f, -7.100^g	-7.260 ± 0.390^b
c_{112}	-0.112	-0.224^a, -0.315^e, -0.267^f, -0.280^g	-0.240 ± 0.040^b
c_{123}	0.130	0.133^a, 0.089^e, 0.055^f, 0.190^g	0.110 ± 0.040^b
c_{144}	0.140	0.127 ± 0.005^a, 0.164^e, 0.174^f, 0.250^g	0.230 ± 0.040^b
c_{166}	-0.258	-0.245 ± 0.002^a, -0.358^e, -0.333^f, -0.310^g	-0.260 ± 0.020^b
c_{456}	0.146	0.118 ± 0.004^a, 0.158^e, 0.164^f, 0.200^g	0.160 ± 0.010^b

[a] see Ref. [25],   [b] see Ref. [42]

[c] see Ref. [53],   [d] see Ref. [45]

[e] see Ref. [23],   [f] see Ref. [27]

[g] see Ref. [48]

so our predicted results of $\gamma(\mathbf{q},\mathbf{j})$ can be serve as a valuable guide or reference for experimenters which would someday execute a measurement.

6. Conclusion

In this work, we have represented a systematic scheme to compute the second- and third-order elastic constants for our researched alkali halide crystals with B1 structure using the DFT and homogeneous deformation method. From the nonlinear fitting, the elastic constants are extracted from a polynomial fit to the energy versus strain data. Our theoretical results for elastic constants are excellent agreement with experimental and previous results. Comparing with linear elasticity theory, our work shows that the third-order effects must be considered when the applied finite deformations are larger than approximately 2.5%. Based on the calculated elastic constants, we have computed the pressure derivatives of SOECs and provided rough estimations for mode Grüneisen constants of long-wavelength acoustic modes. We believe that our ab initio estimates of TOECs can be a very useful tool in applying these alkali halide crystals with B1 structure.

Table 9. Comparison of theoretical and experimental values of the pressure derivatives of the second-order elastic constants (in dimensionless units) for alkali halides

	This work	Previous	Experiment
LiF			
c'_{11}	6.303	9.92^a, 8.221^b, 4.07^c, 11.41^d	
c'_{12}	1.519	2.72^a, 2.916^b, 2.95^c, 2.79^d	
c'_{44}	0.515	1.38^e, 1.38^a, 0.705^b, 0.84^c, 0.45^d	1.38^f
NaF			
c'_{11}	7.868	11.59^a, 11.312^b, 5.26^c, 5.62^d	
c'_{12}	1.210	1.99^a, 3.917^b, 2.57^c, 2.46^d	
c'_{44}	0.021	0.205^e, 0.21^a, 0.031^b, 0.51^c, 0.11^d	0.205^f
KF			
c'_{11}	10.021	6.36^c	
c'_{12}	1.428	2.15^c	
c'_{44}	0.243	0.17^c	
LiCl			
c'_{11}	7.094	7.10^c	
c'_{12}	1.568	2.34^c	
c'_{44}	0.325	0.32^c	
NaCl			
c'_{11}	8.96	12.18^g, 11.71^a, 11.788^b, 8.14^c, 2.98^d	
c'_{12}	1.55	2.25^g, 2.06^a, 1.909^b, 2.21^c, 2.40^d	
c'_{44}	0.24	0.19^g, 0.37^a, 0.139^b, 0.19^c, 0.22^d	0.37^f
KCl			
c'_{11}	10.368	12.52^g, 12.77^a, 12.019^b, 9.29^c, 9.74^d	
c'_{12}	1.068	1.38^g, 1.61^a, 1.924^b, 1.98^c, 2.36^d	
c'_{44}	-0.356	-0.38^g, -0.39^a, -0.415^b, -0.01^c, -0.02^d	-0.39^f

^{a}see Ref. [55], ^{b}see Ref. [47]

^{c}see Ref. [23], ^{d}see Ref. [45]

^{e}see Ref. [44], ^{f}see Ref. [56, 57], ^{g}see Ref. [58]

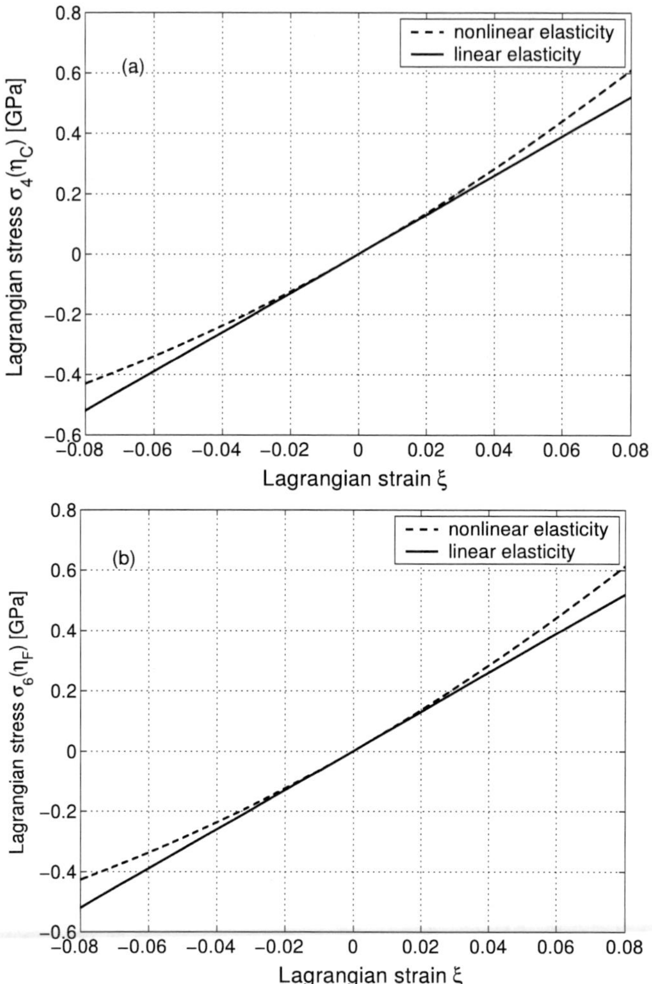

Figure 5. The strain stress relations as a function of linear strain parameter ξ for KCl cubic crystals. Dashed and solid lines represent the curves obtained from nonlinear and linear elasticity theory, respectively. (a) illustrates the stress components σ_4 as a function of linear strain parameter ξ for particular deformation η_C. (b) shows the stress components σ_6 as a function of parameter ξ for the deformation η_F.

Table 10. The phase transition pressure of alkali halide calculated from pressure derivatives of SOECs (in units of GPa)

	This work	Previous	Experiment
LiF	-	55.0^a, 48.0^b	$>10^c$
NaF	-	20.0^c, 12.9^b	$>20^a$
KF	-	5.1^a, 5.1^b	$>10^c$
LiCl	-	16.0^a, 14.4^b	$>10^c$
NaCl	-	7.4^c, 4.9^a, 4.9^b, 28.6^d, 27.0^e, 27.4^f, 23.9^g	29.0^h,30.0^i
KCl	18.2	2.1^a, 2.4^b, 1.1^e, 2.16^m	2.0^s,1.9^t

^{a}see Ref. [60], ^{b}see Ref. [61]
^{c}see Ref. [1], ^{d}see Ref. [62]
^{e}see Ref. [63], ^{f}see Ref. [64]
^{g}see Ref. [65], ^{h}see Ref. [66]
^{i}see Ref. [67], ^{m}see Ref. [68]
^{s}see Ref. [69], ^{t}see Ref. [70]

Table 11. Predictions for the Grüneisen constants $\gamma(\mathbf{q},\mathbf{j})$ of long-wavelength acoustic modes in alkali halides. Theoretical predicition for $\gamma(\mathbf{q},\mathbf{j})$ are calculated on the basis of strain-energy values for c_{ij} and c_{ijk}

	LiF	NaF	KF	LiCl	NaCl	KCl
$q=(\epsilon,0,0)$						
$\gamma(LA)$	1.719	-0.198	2.281	2.006	2.021	2.145
$\gamma(TA)$	0.131	-0.149	0.090	0.047	0.067	-0.608
$q=(\epsilon,\epsilon,0)$						
$\gamma(LA)$	0.936	-0.016	1.553	1.048	1.310	1.397
$\gamma(TA_{xy})$	2.637	-0.617	2.569	2.747	2.343	2.329
$\gamma(TA_z)$	0.131	-0.149	0.090	0.047	0.067	-0.608
$q=(\epsilon,\epsilon,\epsilon)$						
$\gamma(LA)$	0.716	0.049	1.249	0.789	1.033	1.034

References

[1] M. Born and K. Huang, *Dynamical Theory of Crystal Latticess* (Oxford University Press, London, 1956).

[2] D. C. Wallace, *Thermodynamics of Crystals* (Dover, New York, 1998).

[3] F. Birch, *Phys. Rev.* 71 (1947) 809.

[4] F. Murnaghan, *Finite deformation of an elastic solid.* New York: Wiley; 1951.

[5] S. Bhagavantam, *Crystal Symmetry and Physical Properties* (Academic, New York, 1966).

[6] R. Thurston and K. Brugger, *Phys. Rev.* 133 (1964) A1604.

[7] K. Brugger. *Phys. Rev.* 133 (1964) A1611.

[8] Y. Hiki, *Annu. Rev. Mater. Sci.* 11 (1981) 51.

[9] X. D. Zhang, Z. Y. Jiang, B. Zhou, Z. F. Hou and Y. Q. Hou, *Chin. Phys. Lett.* 28 (2011) 076201.

[10] P. N. Keating, *Phys. Rev.* 145 (1966) 637.

[11] P. N. Keating, *Phys. Rev.* 149 (1966) 674.

[12] T. Cağin and J. R. Ray, *Phys. Rev. B* 38 (1988) 7940.

[13] T. Cağin and B. M. Pettitt, *Phys. Rev. B* 39 (1989) 12484.

[14] O. H. Nielsen and R. M. Martin. *Phys. Rev. B* 32 (1985) 3792.

[15] O. H. Nielsen. *Phys. Rev. B* 34 (1986) 5808.

[16] O. H. Nielsen and R. M. Martin. *Phys. Rev. B* 32 (1985) 3780.

[17] M. Łopuszyński , J. A. Majewski. *Phys. Rev. B* 76 (2007) 045202.

[18] J. J. Zhao, J. M. Winey and Y. M. Gupta. *Phys. Rev. B* 75 (2007) 094105.

[19] H. Wang and M. Li. *Phys. Rev. B* 79 (2009) 224102.

[20] R. Wang, S. F. Wang and X. Z. Wu. *Intermetallics* 18 (2010) 1653.

[21] R. Wang, S. F. Wang, X. Z. Wu, Y. Yao and A. P. Liu. *Intermetallics* 18 (2010) 2472.

[22] A. A. Nrayn'yan, *Soviet Phys. Solid State* 5 (1963) 129.

[23] P. B. Ghate, *Phys. Rev.* 139A (1965) 1666.

[24] D. S. Puri, M. P. Verma, *Solid State Commun* 18 (1976) 1295.

[25] Z. P. Chang, *Phys. Rev.* 140 (1965) A1788.

[26] K. D. Swartz, *J. Acoust. Soc. Am.* 41 (1967) 1083.

[27] R. C. Lincoln, K. M. Koliwad, P. B. Ghate, *Phys. Status. Solidi* 18 (1966) 265.

[28] S. Paul, *Ind. J. Pure Appl. Phys.* 8 (1970) 307.

[29] U. Ray, *Solid State Phys.* (India) 29c (1986) 170.

[30] D. C. Wallace, in *Solid State Physics*, edited by F. Seitz and D. Turnbull (Academic, New York, 1970), Vol, 25, p. 301.

[31] R. N. Thurston, in *Physical Acoustics Principles and Methods*, edited by W. P. Mason and R. N. Thurston (Academic, New York, 1964), Vol. 1A, p. 1.

[32] G. Kresse and J. Hafner, *Phys. Rev. B* 48 (1993) 3115.

[33] G. Kresse and J. Furthmüller, *Comput. Mater. Sci.* 6 (1996) 15.

[34] G. Kresse and J. Furthmüller, *Phys. Rev. B* 54 (1996) 11169.

[35] J. P. Perdew, K. Burke, M. Ernzerhof, *Phys. Rev. Lett.* 77 (1996) 3865.

[36] J. P. Perdew, K. Burke, M. Ernzerhof, *Phys. Rev. Lett.* 78 (1996) 1396.

[37] P. E. Blöchl, *Phys. Rev. B* 50 (1994) 17953.

[38] G. Kresse and D. Joubert, *Phys. Rev. B* 59 (1999) 1758.

[39] H. J. Monkhorst and J. D. Pack, *Phys. Rev. B* 13 (1976) 5188.

[40] M. P. Tosi, *Solid State Phys.* 16 (1964) 1.

[41] S. Haussühl, *Z. Phys.* 159 (1960) 223.

[42] J. R. Drabble, R. E. B. Strathen, *Proc. Phys. Soc.* 92 (1967) 1090.

[43] C. V. Briscoe, C. F. Squire, *Phys. Rev.* 106 (1957) 1175.

[44] R. A. Miller, and C. S. Smith, *J. Phys. Chem. Solids*, 25 (1964) 1279.

[45] H. R. Yazar, S. Agan, K. Colakoglu, *Turk J Phys.* 25 (2001) 345.

[46] A. A. Nrayn'yan, *Soviet Physics-Solid State* 7 (1965) 729.

[47] U. Ray, *Phys. B* 405 (2010)3063.

[48] D. Hans, A. A. S. Sangachin, and M. Kumar, *Phys. Stat. Sol.(b)* 147 (1988) 529.

[49] W. A. Bensch, *Phys. Rev. B* 6 (1972) 1504.

[50] J. Vallin, K. Marklund, J. O. Sikström, and O. Beckman, *Arkiv Fysik* 32 (1966) 515.

[51] M. Gluyas, *Brtt. J. Appl. Phys.*, 18 (1967) 913.

[52] J. K. Galt, *Phys. Rev.* 73 (1948) 1460.

[53] S. Yamamoto, O. L. Anderson, *Phys. Chem. Minerals.* 14 (1987) 332.

[54] N. A. W. Holzwarth, G. E. Mattews, R. B. Dunning, A. R. Tackett and Y. Zeng. *Phys. Rev. B* 55 (1997) 2005.

[55] O. L. Anderson, R. C. Liebermann, *Phys, Earth Planet Int* 3 (1970) 61.

[56] R. W. Roberts, C. S. Smith, *J. Phys. Chem. Solids* 31 (619) (1970) 2379.

[57] K. O. Mclean, C. S. Smith, *J. Phys. Chem. Solids* 33 (1972) 279.

[58] D. Lazarus, *Phys. Rev.* 76 (1949) 545.

[59] S. P. Łepkowski, J. A. Majewski and G. Jurczak, *Phys. Rev. B* 72 (2005) 245201.

[60] Y. S. Kim and R. G. Gordon, *Phys. Rev. B* 9 (1974) 3458.

[61] A. J. Cohen and R. G. Gordon, *Phys. Rev. B* 12 (1975) 3228.

[62] H. R. Yazar, *Acta. Phys. Pol. A* 109 (2006) 743.

[63] M. L. Cohen, S. Froyen, *J. Phys. C, Solid State Phys.* 19 (1986) 2623.

[64] R. K. Singh, N. V. K. Prabhakar, *Phys. Status Solidi B* 141 (1987) K29.

[65] K. N. Jog, *PhD. Thesis*, Rani Durgawati University, Jabalpur 1988, p. 77.

[66] D. L. Heinz, R. Jeanloz, *Phys. Rev. B* 30 (1984) 6045.

[67] W. A. Basset, T. Takahashi, H. K. Mao and J. S. Weaver, *J. Appl. Phys.* 39 (1968) 319.

[68] M. P. Tosi, Solid State Phys. 16 (1964) 1; M. P. Tosi and F. G. Fumi, *J. Phys. Chem. Solids* 23 (1962) 359.

[69] P. W. Bridgman, *Proc. Am. Acad. Arts. Sci.* 76 (1945) 9.

[70] G. C. Kennedy, S. N. Vaidya, *J. Phys. Chem. Solids* 32 (1971) 951.

[71] Y. Deng, O. H. Jia, X. R. Chen, J. Zhu, *Phys. B* 392 (2007) 229.

[72] B. B. Karki, G. J. Ackland, *J. Phys. Chem. Solids* 38 (1977) 1355.

[73] A. Mayer and R. Wehner, *Phys. Status Solidi B* 126 (1984) 91.

INDEX

D